Arroyo Center

IMPROVING ARMY PLANNING FOR FUTURE MULTINATIONAL COALITION OPERATIONS

Thomas S. Szayna, Frances M. Lussier, Krista Magras, Olga Oliker, Michele Zanini, Robert Howe

RAND

Prepared for the
United States Army

Approved for public release; distribution unlimited

The research described in this report was sponsored by the United States Army under Contract No. DASW01-96-C-0004.

Library of Congress Cataloging-in-Publication Data

Improving Army planning for future multinational coalition operations / Thomas S. Szayna ... [et al.].
p. cm.
"MR-1291-A."
Includes bibliographical references.
ISBN 0-8330-2960-6
1. Combined operations (Military science) 2. United States. Army. 3. Military planning—United States. I. Szayna, Thomas S., [DATE]

U260 .I47 2001
355.4'6—dc21

00-067353

Published 2001 by RAND
1700 Main Street, P.O. Box 2138, Santa Monica, CA 90407-2138
1200 South Hayes Street, Arlington, VA 22202-5050
201 North Craig Street, Suite 102, Pittsburgh, PA 15213-1516
RAND URL: http://www.rand.org/
To order RAND documents or to obtain additional information, contact Distribution Services: Telephone: (310) 451-7002; Fax: (310) 451-6915; Internet: order@rand.org

PREFACE

This report documents the results of a project entitled "Improving Ground Force Performance in Future Coalitions." The project aimed to improve the Army's planning of its activities designed to enhance the operational performance of ground forces in coalition operations across the spectrum of missions.

The research reported here was sponsored by the Military Deputy to the Deputy Under Secretary of the Army for International Affairs (DUSA-IA) and was conducted in RAND Arroyo Center's Strategy, Doctrine, and Resources Program. The Arroyo Center is a federally funded research and development center sponsored by the United States Army.

The report should be of interest and use to those concerned with the international activities of the Army and specifically with ways of enhancing its compatibility with potential ground force coalition partners. The report has most relevance to the Army, but it should also be of use to CINCs and the Joint Staff.

For more information on RAND Arroyo Center, contact the Director of Operations (telephone 310-393-0411, extension 6500; FAX 310-451-6952; e-mail donnab@rand.org), or visit the Arroyo Center's Web site at http://www.rand.org/organization/ard/.

CONTENTS

FIGURES

TABLES

SUMMARY

The Army Vision recognizes explicitly that in future operations, Army forces will perform missions as part of a larger joint-combined-multinational force. Given the importance that Army doctrine places on coalition operations, and the capability gap between the U.S. Army and even the most sophisticated partner ground forces, it will be critical to take steps to increase multinational force compatibility (MFC) before any deployment. Preparing for such a future is one aspect of the Army's Title 10 functions, and it entails finding ways to enhance the long-term compatibility of its units with units of the most important and most likely coalition partners. The Army's international activities (IA), most of which enhance MFC, are the main vehicle for meeting this goal. This study's principal purpose is to help improve the Army's planning of its IA activities, in order to enhance the performance of ground forces in coalition operations across the spectrum of missions. This entails two key steps: (1) improve the organizational mechanisms to monitor and administer Army international activities, and (2) devise a long-term MFC plan.

IMPROVING ORGANIZATIONAL MECHANISMS

In recognition of the need for greater coordination of its international activities, the Army set up the office of the Deputy Under Secretary of the Army, International Affairs (DUSA-IA), with the mandate to plan, coordinate, and facilitate the Army's international activities. Despite the creation of DUSA-IA, a coherent formulation and implementation of the Army's IA is still hampered by a lack of information sharing and coordination across Army agencies. Such coordination remains elusive because the current budgetary system makes it difficult to identify resources devoted to IA. Moreover, DUSA-IA/MFC does not have the ability to monitor or influence the allotment of funds for all relevant Army international activities programs. This leads to two consequences detrimental to Army MFC efforts. First, the lack of clear and detailed knowledge of resources devoted to the Army's international activities (and MFC efforts specifically) reduces the visibility of the Army's efforts in intra-Army planning, programming, and budgeting process. Second, the lack of knowledge of the resources devoted to MFC makes it virtually impossible to conduct an effective evaluation of the Army's MFC efforts, thus constraining severely the Army's ability to assess tradeoffs in MFC efforts and to choose MFC efforts accurately. Thus, the visibility and verifiability requirements of Army MFC efforts (necessary in view of the explicit importance placed on multinational coalition operations in the Army Vision) are not being met.

The root causes of the problem lie in the lack of effective and appropriate planning processes, where the effectiveness of coalition forces as a whole is increasingly important to mission success. The current system was not designed for, and therefore is not capable of, supporting DUSA-IA in its official role as the agency that (1) establishes and disseminates policy guidance on how Army IA programs set substantive objectives and funding priorities, and (2) provides a high-level evaluation of how the MFC "system" is performing as a whole. This in turn makes it difficult for DUSA-IA to fulfill its responsibilities.

Effective management of the Army's resources for international activities will require changes in the collection and analysis of data on funding for such activities. Under the current system, the lack of information on funding or activities by individual country makes it difficult for the Army to know whether activities with some countries are under- or overemphasized. It is possible that the on-line data base, the Theater Engagement Plan Management Information System (TEPMIS), will be a mechanism to do this, since TEPMIS is envisioned to provide consolidated data on Army international activities of interest to the CINCs. This would cover most of the activities that contribute significantly to enhanced MFC capabilities, such as exercises, officer exchanges, and the IMET program. Resources devoted to activities that may not be included in TEPMIS, such as those for cooperative R&D, are easily identified within the Army's budget. Combining such data with that from TEPMIS should provide a relatively complete picture of the distribution of resources for Army IA—helping the CINCs, the Army agencies implementing the various programs, and DUSA-IA in its role as proponent of MFC-enhancing activities.

In seeking to improve DUSA-IA's functioning, the Army might want to consider several alternative structures for managing the execution of its IA programs. We recommend a strengthening of the two-way relationship between DUSA-IA and the field offices, a link necessary to make DUSA-IA the key advocate and center of the establishment and dissemination of policy guidance on how IA programs should be run. In this case, DUSA-IA would not control funding for all IA, but it would have visibility of and influence over the allocation of funding for IA, particularly those that directly enhance MFC.

DEVISING A LONG-TERM MFC PLAN

For DUSA-IA to be effective in its role as the main proponent of the Army's MFC-related activities, it needs to have an overall blueprint or vision of what it needs to accomplish. In the area of prioritization of partners and activities, current Army MFC efforts are not guided by any concrete long-term plan. Yet without such a blueprint, there is a real danger of ineffective and inefficient outcomes or, at a minimum, a lack of synergy between the various Army activities designed to enhance MFC. We propose a four-step process that (1) identifies the most likely long-term U.S. coalition partners, (2) provides a means of pinpointing the compatibility shortcomings of the potential partners across the full range of missions, (3) links specific Army MFC policies to the shortcomings, and (4) allows for cost-efficiency assessments at the program level and still within the overall planning framework, so as to enable the determination of cost-effective resource allocation. Once cost-effectiveness assessments of MFC efforts become possible, a fifth—currently hypothetical—step of integrating the Army's own force planning with that of allies and likely partners would become a realistic option. In other words, Army planners could carry out cost-benefit assessments on the basis of tradeoffs between own and ally capabilities, knowing in detail the costs involved and possibly in cooperation with select allies and partners.

To identify the most likely U.S. partners, we developed a methodology for anticipating the propensity of a given country to join the United States in a coalition operation. We applied the methodology to all countries of the world with a military establishment. To identify compatibility shortcomings, we developed the Military Compatibility Assessment Tool (MCAT), a substantively modified version of RAND's DynaRank decision support system (and a Microsoft Excel workbook available for Macintosh and Windows-based personal computers) designed at RAND in 1998 to assist Department of Defense decisionmaking. Both the methodology for identifying countries likely to participate with the United States in coalition operations and the MCAT can be used by the Army in its future evaluations. The methodologies developed here are not meant to be used mechanistically. But they do provide shortcuts to decisionmaking and assist in creating a cohesive overall Army MFC policy.

ACKNOWLEDGMENTS

The authors are grateful to MG Howard J. von Kaenel, the Military Deputy to the Deputy Under Secretary of the Army (International Affairs) for sponsoring the research. We gratefully acknowledge the interest and support from the staff at DUSA-IA, especially Lawrence E. Ewert (chief, Multinational Force Compatibility Branch), Donald R. Davidson, Mark E. McDonough, Jeff Stefani, and Andy Tarr.

Many people in the Department of the Army, the Joint Staff, and the Office of the Secretary of Defense deserve thanks for their time and cooperation with the project staff. The following individuals were especially helpful in the course of this research: COL Karl Farris (ret.) at the U.S. Army Peacekeeping Institute, LTC Richard R. Caniglia at the American, British, Canadian, Australian Armies' Standardization Program (ABCA), and Mr. Jim Stone at the Joint Warfighting Center (JWFC). Harry F. Rowley (DUSA-IA), Elizabeth A. May (DUSA-IA), and Craig College (Deputy Director of PA&E) shared their knowledge of and provided data on Army resources.

The authors are indebted to Paul Davis (RAND), COL Karl Farris (ret.), and Thomas Young (Naval Postgraduate School) for providing thoughtful comments on an earlier draft of this report. We also thank Kevin McCarthy and Jefferson Marquis of RAND, as their work on engagement strategy influenced our views toward MFC. Nikki Shacklett edited the report.

ACRONYMS

ABCA	American, British, Canadian, and Australian Armies' Standardization Program
ACRI	African Crisis Response Initiative
AIA	Army International Activities
AIAP	Army International Activities Plan
AMC	Army Materiel Command
ANZUS	Australia, New Zealand, and United States
ARF	ASEAN Regional Forum
ARRC	Allied Command Europe (ACE) Rapid Reaction Corps
ASEAN	Association of Southeast Asian Nations
C4I	Command, Control, Communications, Computers, and Intelligence
CAA	Capability Assessment Area
CIA	Central Intelligence Agency
CINC	Commander in Chief (commander of a combatant command)
CIS	Commonwealth of Independent States
CM	Compatibility Measure
CNA	Center for Naval Analyses

CONUS	Continental United States
CTR	Cooperative Threat Reduction
DCSLOG	Deputy Chief of Staff, Logistics
DoD	Department of Defense
DUSA-IA	Deputy Under Secretary of the Army, International Affairs
ECOMOG	Economic Community (of West African States) Monitoring Group
EU	European Union
FAO	Foreign Area Officer
FSU	Former Soviet Union
GUUAM	Georgia, Ukraine, Uzbekistan, Azerbaijan, and Moldova
HNS	Host Nation Support
HQDA	Headquarters, Department of the Army
IFOR	NATO Implementation Force
IMET	International Military Education and Training
INF	Intermediate Range Nuclear Forces
JCS	Joint Chiefs of Staff
KFOR	NATO Kosovo Force
LMI	Logistics Management Institute
MACOM	Major Army Command
MCAT	Military Compatibility Assessment Tool
MDEP	Management Decision Package
MFC	Multinational Force Compatibility
MOD	Ministry of Defense
MOU	Memorandum of Understanding
NATO	North Atlantic Treaty Organization

O&M	Operations and Maintenance
OAS	Organization of American States
OAU	Organization of African Unity
OSD	Office of the Secretary of Defense
PE	Program Element
PEG	Program Evaluation Group
PEO	Program Evaluation Officer
PfP	Partnership for Peace
PLA	People's Liberation Army, Republic of China
PM	Political-Military
RDT&E	Research, Development, Test, and Evaluation
RSI	Rationalization, Standardization, and Interoperability
RSO&I	Reception, Staging, Onward Movement, and Integration
SA	Security Assistance
SAO	Security Assistance Officers
SATCOM	Satellite Communications
SEATO	South East Asia Treaty Organization
SFOR	NATO Stabilization Force
SIGINT	Signal Intelligence
SINCGARS	Single Channel Ground and Airborne Radio System
SOFA	Status of Forces Agreement
TDA	Table of Distribution and Allowances
TEPMIS	Theater Engagement Plan Management Information System
TOE	Table of Organization and Equipment
TRADOC	U.S. Army Training and Doctrine Command

TSPS	Theater Security Planning System
USMA	United States Military Academy
WEU	Western European Union

Chapter One

INTRODUCTION

THE CONTEXT

Each year, U.S. Army personnel participate in thousands of activities that include personnel from the armed forces of other countries. These activities touch on virtually every aspect of the Army, ranging from cooperation in research and development, to joint training and exercises, to staff talks with visiting foreign military delegations. While the specific rationale for undertaking these activities may stem from meeting U.S. alliance commitments or from other agreements on cooperation and exchange, the Army's international activities collectively represent a critical part of the Army's input to and everyday implementation of the national policy of "engagement."

The participation of Army personnel in myriad international activities also has more specific military ramifications of interest to the Department of the Army, especially in preparing for potential coalition operations. Whatever their rationale, many if not most of the Army's international activities contribute in one way or another to improving the performance of ground forces in potential coalition operations involving the United States. Some activities, such as those leading to equipment standardization or the adoption of similar doctrine, have a clear link to enhancing compatibility. Other activities may have a more indirect link, such as providing a measure of personal acquaintance and trust between U.S. and foreign military personnel, which may or may not become significant for coalition purposes.

The Army has had an enduring interest in improving the effectiveness of ground-force coalition operations, but that interest has grown appreciably and changed in focus during the past decade. Previously, the Army's interest in coalition operations focused primarily on achieving effectiveness with allies in high-intensity combat operations. Currently, the Army faces the challenge of having to prepare for coalition operations that span the full spectrum of military missions (peacekeeping to peace enforcement to conventional combat), in conditions of uncertainty about coalition partners (countries ranging from established allies to ad hoc partners to erstwhile adversaries), and against the background of a growing gap in military capabilities between the United States and even its closest and most affluent allies. Amidst this more difficult and uncertain environment for planning effective coalition operations, the reductions in U.S. force size in the 1990s have increased the probable role that allies and partners will play in future coalitions.

Most of the major military operations that the United States has undertaken since World War II have included coalition partners and allies. The trend toward greater participation in multinational coalition operations has become even more pronounced during the past decade, especially for ground force deployments. Although the United States will retain the ability to act unilaterally, its fundamental security policy favors securing goals through multinational coalitions when possible. Thus, it appears almost certain that most future U.S. military operations will take the form of coalition operations. Regardless of whether the participation of other coalition partners ranges from little more than a political "fig leaf," to a facilitator, to a major force contributor, the effective integration of U.S. and coalition partner forces to accomplish mission goals will be crucial to the success of future multinational operations.[1]

Given such an outlook, improving the operational performance of ground force coalition operations stands among the most important tasks for the Army. Collectively, the activities designed to achieve that goal fall under the general heading of "enhancing MFC," with MFC defined as "the ability of the United States Army and its forces

---

[1]Future coalition operations will also entail coordination between U.S. Army units and international organizations, such as UN agencies, as well as with nongovernmental organizations. Although these linkages are important for the overall success of an operation, they are not considered here. The focus of this report is on enhancing MFC in the military-to-military domain.

to operate effectively as an ally or coalition partner across the full spectrum of military missions."[2] MFC enhancement activities touch on all aspects of force development. The term MFC contrasts with the concept of rationalization, standardization, and interoperability (RSI) previously used by the Army. Rather than focusing solely on materiel-technical issues, as implied by RSI, MFC also includes training, doctrine, organization, and procedures. MFC efforts have the overarching goal of achieving compatibility with partner armies in a broad set of operational categories, and, as such, they represent a holistic approach to improving the performance of multinational coalition operations.

OBJECTIVES, APPROACH, AND ORGANIZATION

In an overall sense, this study seeks to help the U.S. Army improve its processes for the development, guidance, and evaluation of MFC policies. It does so by helping the Army answer the following key sets of questions about the Army's MFC policy:

- When designing its coalition enhancement activities, how should the Army select among the variety of allies and partners? Which partners should be the primary targets for MFC efforts? Should some operations on the mission spectrum receive greater attention than others?
- To achieve more effective coalition operations, how can the Army identify the most pressing problems in other armed forces? What programs are available to deal with these problems? What is the best way to evaluate their effectiveness?
- How should the Army integrate its MFC efforts with its force development processes? In other words, how can the Army ensure that increasing the effectiveness of allied and partner armies

---

[2]This definition is taken from *Multinational Force Compatibility: Central Concepts*, briefing by DUSA-IA, August 1998. The definition builds on the concepts of "multinational force" and "compatibility," as defined in the Department of Defense Dictionary of Military Terms. Multinational Force: "A force composed of military elements of nations who have formed a temporary alliance for some specific purpose." Compatibility: "Capability of two or more items or components of equipment or material to exist or function in the same system or environment without mutual interference." *http://www.dtic.mil/doctrine/jel/doddict/*.

in coalition operations with the United States enhances the Army's decisionmaking processes for its own force structure?

To tackle these questions, we began with a comprehensive study of the existing state of affairs with regard to the Army's formulation and implementation of MFC policy. We examined the issue of resources and funding of the Army's international activities. We also examined the evaluation processes for MFC activities and Army objectives for MFC, based on guidance that the Army receives from the Department of Defense and the Joint Staff. We found that standard cost-benefit analyses of the Army's MFC efforts are currently difficult, if not impossible, because of data unavailability and a lack of overall Army guidelines that would allow for an evaluation of MFC efforts. Chapter Three documents our efforts.

To provide the framework that would allow cost-benefit assessments of MFC efforts in the future, we then formulated an alternative approach to MFC policy. Using current scholarship on alliances and coalitions, we devised a methodology for setting priorities in the Army's MFC policy. We also adapted a RAND-developed decision support system (DynaRank) to allow it to assess the extent and scope of other armies' compatibility shortfalls vis-à-vis the U.S. Army. The new tool, the Military Compatibility Assessment Tool (MCAT), points to the main compatibility problems and matches the appropriate policies and programs to ameliorate the problems. In conjunction with changes in the availability and organization of data on resources spent on MFC, further modifications of DynaRank will allow for an evaluation of the effectiveness of the policies and, potentially, a consideration of allied and partner capabilities within the Army's own force development process. Chapter Four outlines our proposed alternative approach. Three appendixes provide a database for the Army to use in MFC policy planning, documentation of the application of our methodology, and a tutorial for using the MCAT.

The initial assessment of current MFC policy is based on a review of all relevant documents and regulations. We also interviewed numerous Army and Department of Defense (DoD) personnel working with MFC policy and on MFC issues. In addition, we conducted a literature search for other analytical efforts on the topic of MFC. In proposing an alternative approach to MFC policy, we borrowed heavily from relevant political science, and especially international relations, theory. The MCAT stems from recent operations research work at RAND.

Chapter Two

THE PROBLEM

The Army Vision makes it clear that operating in a coalition framework has become an assumption underlying the future employment of Army forces:

> The spectrum of likely operations describes a need for land forces in joint, combined, and multinational formations for a variety of missions extending from humanitarian assistance and disaster relief to peacekeeping and peacemaking to major theater wars.[1]

In this sense, the current vision does not differ from the projections in Army Vision 2010:

> Land component operations in 2010 will be fully integrated with those of joint, multinational, and non-governmental partners. Recent experience reminds us that Army operations have never been and will never be independent.[2]

The Army's official doctrine recognizes that multinational operations can be beneficial: (1) politically, because they give a particular deployment greater legitimacy, and (2) militarily, when effective coalitions relieve the United States of the burden of supplying most of the troops and assets.

The continuing importance of MFC operations has implications for how the Army approaches international activities in peacetime.

---

[1]The Army Vision. *http://www.army.mil/armyvision/armyvis.htm.*

[2]Army Vision 2010. *http://www.army.mil/2010.*

Operating effectively and to the fullest potential in a multinational coalition framework entails a successful effort by all participants to think, plan, and resource as coalition partners. While ad hoc "workarounds" just before an operation may result in at least minimum effectiveness in the given coalition mission, such arrangements have little chance of enabling the combined coalition forces to reach their full potential. A better course of action that genuinely addresses the long-term Army vision must include extensive peacetime cooperation, especially with the most important and most likely partners. The Army's international activities, most of which enhance MFC, thus play a vital role in implementing the Army's long-term vision.

In theory, Army MFC efforts derive from the following flow of decisions:

1. National-level (President's guidance) and Office of the Secretary of Defense (OSD) guidelines provide the strategic parameters for the Army's international activities, including its MFC efforts.

2. In cooperation with the Joint Staff and the commanders in chief (CINCs), Army planners prioritize partners and identify their major problems with conducting coalition operations with Army units.

3. Army planners and the CINCs define the appropriate MFC packages and allocate resources.

4. Army component commands and agencies implement the plans and provide feedback on the effectiveness of the efforts. Feedback loops also provide a check on the continued appropriateness of the MFC packages and prioritization choices.

The Army has a secondary role in steps 2 and 3, and a central role in step 4. But in practice, rather than being secondary, the institutional Army's role in steps 2 and 3 is negligible, and no recognizable Army perspective is brought into the process. This is not an ideal outcome, because the institutional Army, as part of its "train, organize, and equip" function of preparing the force, has a clear interest in advancing compatibility with the likely coalition partner forces. The primary mechanism to achieve greater compatibility is the Theater Engagement Plan (TEP), prepared by each CINC. The TEP amounts to a list of all the envisioned activities with individual militaries in the given theater. Each CINC's preferences as to activities and partners are driven by a variety of military and nonmilitary factors. From the Army's perspective of preparing for coalition operations, some of the military factors behind CINC preferences may not be optimal. For example, the preferences may reflect a CINC's dominant short-term or, at best, mid-term, focus, or they may not take sufficiently into account the coalitional potential of some countries' forces in operations outside the given CINC's command. If the Army component commands were to act as proponents of the Army's perspective within the process of constructing a TEP, then they would ensure that longer-term institutional Army concerns are taken into account. Moreover, such a role has the potential to advance the national-level goal of achieving better ground force compatibility with the likely coalition partners in future operations while avoiding the artificial constraints of CINC commands or the limited time-frames set by the CINCs.

For Army component commands to act as proponents of the Army's perspective on MFC within the TEP process, they need an overall guiding Army blueprint for MFC, one that the Army component commands could also use in their own planning process. Moreover, to come up with such a general MFC blueprint, the Army needs a way to determine which resources are for advancing MFC and to evaluate the efficiency and effectiveness of MFC efforts. The problem is that even though the Army has recognized the need for greater coordination of its international activities, its current accounting and information systems for tracking MFC activities are insufficiently developed.

None of the above is to question the dominant role of the CINCs in planning international activities within their commands. Instead, the suggestion that the Army can and should play a stronger role in the TEP process is meant to strengthen the ability of the services and the combatant commands to achieve national-level goals for improving the coalition readiness of U.S. forces.

In recognition of the need for greater coordination of the Army's international activities, in 1995 the Secretary of the Army set up the office of DUSA-IA to plan, coordinate, and facilitate the Army's international activities.[3] Despite the creation of DUSA-IA, a coherent formulation of guidelines for the Army's international activities is still hampered by a lack of information sharing and coordination across Army agencies. Such coordination remains elusive because (1) the current budgetary system makes it difficult to identify resources devoted to IA, and (2) DUSA-IA/MFC does not have the ability to monitor or influence the allotment of funds for all relevant Army international activities programs. The lack of information sharing and coordination leads to two consequences detrimental to Army MFC efforts. First, the lack of clear and detailed knowledge of resources devoted to the Army's international activities (and MFC efforts specifically) reduces the visibility of the Army's efforts in the intra-Army planning, programming, and budgeting process. With few funds under its control, DUSA-IA faces the problem of being relegated to a level of little importance and sometimes even "below the decimal line" in intra-Army decisionmaking. Second, the lack of knowledge about the resources devoted to MFC makes it virtually impossible to evaluate meaningfully the Army's MFC efforts. Inability to calculate the cost-effectiveness of MFC efforts makes it difficult to present the potential benefits in terms of resources and thus nearly impossible to assess tradeoffs accurately and choose MFC efforts judiciously. For instance, a lack of information on funding or activities by individual country makes it difficult for the Army to know whether activities with some countries are under- or overemphasized. Thus, the visibility and verifiability that the Army's MFC efforts require in view of the explicit importance placed on multinational coalition operations in the current Army Vision (as well as in Army Vision 2010) are not provided. This is a problem of an absence of effective and appropriate processes and a lack of Army adaptation to contemporary planning, where the effectiveness of coalition forces as a whole is increasingly important to the success of a mission. The current system was not designed for, and therefore is not capable of, supporting DUSA-IA in its official role as the agency that (1) establishes and disseminates policy guidance on how Army IA programs are to set substantive objectives and funding priorities

---

[3]The responsibilities of DUSA-IA were codified in 1997: General Orders No. 10, *Assignment of Functions, Responsibilities, and Duties Within the Army Secretariat*, Washington, D.C.: Headquarters, Department of the Army, August 12, 1997.

and (2) provides a high-level evaluation of how the MFC "system" is performing as a whole.

For the reasons mentioned above, it is no easy matter to put together an analysis of the Army's MFC efforts along the lines of either efficiency or effectiveness. Indeed, standard evaluations of the Army's MFC efforts meet currently impregnable data barriers.

In terms of an efficiency, or cost-benefit, evaluation, the way the Army organizes its budgetary data makes it extremely difficult to calculate the inputs (measured in funds devoted to specific activities) needed to arrive at some measure of the costs of international activities in general, and activities directly influencing MFC specifically. Data that would allow for calculations of specific figures for Army MFC activities organized along functional and geographic lines are currently not available to RAND.

In terms of the outputs of any cost-benefit evaluation, measures of the compatibility levels of allied and partner militaries in crucial categories are available, though they are not organized in an easily attainable form. The best assessments of the compatibility of other militaries come from the Army's own experiences with them. Both the Joint Staff and the Army gather comprehensive data on other militaries and lessons learned from multinational exercises and operations. However, evaluation processes capture MFC lessons learned only in the context of the specific operation and the relationship of the MFC issues to the commander's operational objectives. In addition, the evaluation of program effectiveness at the level of individual programs or exercises is not complemented by a robust "macro" view of whether resources are being allocated effectively for the whole system.

Any evaluation of the effectiveness of MFC efforts has to compare the outcomes of those efforts against the goals they are meant to accomplish. But the Army has no identifiable institutional set of goals in the MFC realm. Individual Army commands and agencies set priorities for international and MFC efforts, but any stated goals and benchmarks are program-specific. Summing them does not necessarily translate to an Army set of goals.

The current situation makes it difficult to identify the shortcomings that may be present in existing MFC efforts and to provide a fully justifiable set of recommendations for improvement. The remainder of this report deals with the two basic problems, the transparency of resource data and the formulation of a central guiding plan.

Chapter Three presents a detailed overview of how budgetary data are now compiled, and it outlines some recommendations to change the current state of affairs. As for a set of Army guidelines regarding MFC, we tackled the problem backwards. Thinking along the lines of the desired "end state," we constructed a framework for thinking about Army MFC policy that focuses on what the Army, within its institutional constraints, ideally would wish to accomplish in the realm of MFC. Then we worked backward to identify the specific policies that might lead to such an ideal end state. Most of all, defining the desired end state requires accurate identification and ordering of partners for cooperation in MFC efforts. Among the partners identified as priorities, the Army then needs to ascertain areas of need and match appropriate measures to deal with them. This approach has the merit of outlining the most appropriate directions for Army MFC efforts even without a detailed knowledge of the efficiency and effectiveness of current efforts. The framework and the decision support system that we propose have utility for long-term planning of Army MFC policy even under the current conditions of little knowledge of resources spent on MFC. The framework we propose tackles the fundamental question behind the Army's MFC efforts, namely, how to make sure that the Army cooperates with the proper partners in an appropriate manner.

Chapter Three

RESOURCES FOR IMPROVING MFC EFFORTS

The allocation of resources is a critical aspect of managing the Army's international activities in order to enhance MFC. Good management of resources requires knowing where they are invested and what kind of return they are generating. This chapter focuses on identifying where the Army invests its funds for international activities. Chapter Four will address the question of how to maximize their yield.

Managers need to have an appreciation of the resources that might be available to them, which in this case means a knowledge of the total investment allotted to international activities by the Army as well as those funds earmarked for MFC. Such information, coupled with a sense of the marginal investment needed to increase the level of some activities, would enable managers to reallocate funds among the Army's international activities if some programs proved more beneficial than others. In short, knowing what resources are available for MFC activities, how they are allocated among individual programs, and what the marginal cost is for each event within a program (what an additional annual exercise would cost, for example) are prerequisites for good management of the Army's efforts to enhance MFC.

DIFFICULTY IN IDENTIFYING RESOURCES DEVOTED TO ENHANCING MFC

Identifying Army resources devoted to activities that enhance MFC is difficult, and the underlying reasons are myriad and varied.[1] One problem is that not all of the Army's international activities directly benefit MFC. For example, the Army's efforts in implementing the INF treaty are obviously an international activity, but it would be difficult to identify the benefits to MFC that would result directly from such an activity. Another problem is that the Army's international activities incur indirect costs that are very difficult to capture. For example, the travel costs incurred by Army personnel when they attend bilateral and multilateral meetings overseas can be identified as direct costs of IA. But unless the attendees are involved in IA full time, it is likely that the time they spend attending the meeting—that is, a portion of their salaries—is an indirect cost associated with the meeting that cannot be easily captured.

But the broadest reason it is difficult to determine the level of resources devoted to Army international activities is that such activities are pervasive throughout the Army's operations. Although the integration of the Army's international activities into its everyday tasks benefits and furthers the Army's strategy of engagement, it makes it difficult to isolate those aspects of Army operations that constitute IA. As a consequence, direct funding for those activities is spread across congressional appropriations accounts and throughout the Army's internal budget management structure. This led a previous study of Army international activities to state that "Army elements are participating in some international programs that are not recognized as such, and total Army personnel and financial resources being expended for international programs are unknown."[2]

---

[1]Except where noted, all discussion of funds or resources refers to amounts requested of Congress, appropriated by Congress, or included in Army budget planning documents. It does not refer to the amount of funds actually spent.

[2]David Kaplan, James Keenan, James McKimmey, and George Sinks, *Improving Policy Formulation and Management of Army International Activities*, McLean, VA: Logistics Management Institute, AR505MR1, June 1996, p. 8.

ARMY BUDGET STRUCTURE

The Army's annual budget, which totaled $69 billion in fiscal year 1999, is divided up different ways for different purposes. The Congress appropriates funds annually for the Army in six major accounts (Military Personnel, Operations and Maintenance, Procurement, RDT&E, Family Housing, and Military Construction). Within each of these accounts, program elements (PEs) fund specific programs. The Army manages distribution of its budget internally using a totally different structure based on major functions, such as training and equipping the Army. Funds for these functions are aggregated into program element groups, which are broken into smaller management decision packages. Neither budget structure lends itself to easy identification of funding for IA.

The detail included in public documents describing the Army's budget based on the appropriated account structure is often insufficient for determining IA funding. Most of the funds appropriated for IA are in the Operations and Maintenance (O&M) account. This account provides funds for the operating costs of activities such as major exercises, training programs, overseas schools such as the Marshall Center, and personnel exchanges and visits. Nevertheless, budget documents do not specify O&M costs to the level of detail needed to identify funding for many individual IA programs. Similar problems exist with respect to the cooperative research and development (R&D) efforts that are funded out of the RDT&E account. Many international R&D efforts are subprojects within larger R&D efforts, with separate funding not identified. Finally, there is no detail in public budget documents on the military pay account that would identify funds for soldiers, such as foreign area officers, who are involved full time in international activities. Thus, publicly available budget documents provide, at best, an incomplete picture of funds associated with IA.

The Army's internal budget management structure is equally ill suited for identifying resources devoted to IA. The Army's budget is managed by a small number of Program Evaluation Groups (PEGs) and is divided into a much larger number of Management Decision

Packages (MDEPs). Out of the more than 550 MDEPs, about 50 have been identified as including funds for IA.[3]

International activities for the most part do not fit neatly into either of these budget systems (see Table 3.1). This may be because they were designed long before IA, as such, became an important part of the Army's strategy. As a consequence, funds for some activities (for example, exercises with forces from other countries) might be found in more than one MDEP in the Army's internal system and more than one PE in the President's Budget documents. This makes it dif-

Table 3.1

Comparison of Budget Structures

Internal Army Budget			President's Budget	
PEG	MDEP	Activity	PE Number[a]	Account
Training	VJCS	Joint Exercises (CJCS)	121011	O&M
Training	JDHB	Developing Country	442010	O&M
		Combined Exercise Program	131079	O&M
Training	TAMC	The Marshall Center	131096	O&M
			132078	O&M
			442010	O&M
			and others	O&M
Training	VRSI	Standardization and	441004	O&M
		Interoperability	442015	O&M
			665801	RDT&E
			665802	RDT&E
Sustaining	VFMS	FMS, Unreimbursed Costs	442010	O&M
			442015	O&M
			and others	O&M

NOTE: Budget data provided at subactivity level only, not individual PEs.

[a]Individual MDEPs generally do not include all funds of specified PEs.

[3]The Army maintains the PROBE budget database, which is updated periodically throughout the year, to support the Army's budgeting process and to reconcile its appropriated budget—which is broken into six accounts, each with numerous program elements or PEs—with its own internal budget management system of PEGs and MDEPs.

ficult to manage resources for such activities within the Army and justify the need for them before Congress.

A final obstacle to identifying resources allotted to IA is that some are not included in the Army's budget at all. Some security assistance programs, such as foreign military sales (FMS), are paid for by fees collected on the sale of U.S. manufactured equipment to foreign governments. Other funds are appropriated to the Defense Department or the State Department and then transferred to the Army.[4] Although such activities do not require resources from the Army's budget, they do often demand the time of Army military personnel and civilians. Since there is a ceiling on the number of these personnel that the Army may retain, even activities that demand little or no funds from the Army's budget place demands on limited Army resources.

PREVIOUS ATTEMPTS TO IDENTIFY RESOURCES DEVOTED TO IA

Three studies in the past 10 years have addressed the issue of management and funding of international activities within the services.[5] The Center for Naval Analyses (CNA) looked at the Navy's international activities in a study published in 1998. Although CNA did not try to quantify the level of resources devoted to Navy IA, it did try to identify the sources of those resources. The study concluded that funding for the Navy's international activities came from many different sources, a finding that was common to all three studies. The study of Army international activities by the Logistics Management Institute (LMI) identified three main sources of funding for Army international activities, but it was not able to provide a detailed funding profile for individual activities or an estimate of total funding. The most extensive attempt to identify the Army's international activities and the resources devoted to them was conducted by Calibre Systems in the mid-1990s. This study also concluded that it

---

[4]The International Military Education and Training program is an example of a program funded this way.

[5]Jonathan D. Geithner, Margaret Daly Hayes, Catherine K. Lea, Marvin A. Pokrant, Patrick Roth, *OPNAV International Activities: Who's Doing What With Whom*, Alexandria, VA: Center for Naval Analyses, CRM 98-107.10, October 1998; Kaplan et al.; and "Army International Programs," Falls Church, VA: Calibre Systems, Inc., undated draft.

is not possible to identify all the resources devoted to Army international activities using standard Army databases.

LMI Study

The LMI study acknowledged the importance of identifying resources devoted to IA and attempted to quantify the largest sources for them. LMI divided Army international activities into three groups based, to some extent, on the source of funds for those activities. The three categories, which DUSA-IA has retained to describe the Army's international activities, were security assistance, political-military, and materiel-technology. The first group, security assistance (SA), is, by some measures, the largest in financial terms. It includes foreign military sales, international military education and training (IMET), the transfer of excess defense articles, and several other activities. In a given year, the Army might oversee billions of dollars worth of transactions in the SA programs. LMI identified almost 2,000 man-years of military and civilian personnel effort devoted to planning, managing, and executing SA activities. Most of these personnel costs, however, are paid through direct charges levied on foreign customers and are not funded by or included in the Army's budget.

The Army's political-military (PM) activities include a wide array of "diverse, unstructured, and independent" programs, according to LMI.[6] This category encompasses major exercises, bilateral and multilateral meetings, and international visits by foreign delegations, attendance at foreign schools, personnel exchanges, and reserve component activities. LMI concluded that the Army's identifiable expenditures for these manpower-intensive activities, which they estimated at $300 million per year, captured only a small portion of the entire cost.

LMI described the last category, materiel-technology (M-T), as being the smallest in scope and resources. LMI concluded, however, that the programs in this category, including cooperative logistics support, foreign military purchases, and cooperative research and development programs, held the potential for cost savings and en-

---

[6]Kaplan et al., p. 2.

hanced interoperability for the Army with the small number of highly developed countries involved. LMI did not estimate the amount of resources invested in this category, stating only that it varied widely from year to year and that it was small compared to the amount devoted to security assistance and political-military activities.

Calibre Study

Calibre Systems was enlisted by the Army in the mid-1990s to identify the dollar resources associated with its international activities programs. Of the 141 international activities identified in the 1991 Army International Activities Plan (AIAP), however, Calibre Systems could identify the level of annual funding for only 16 (see Table 3.2).

Total annual funding for all 16 activities reached almost $1 billion, while annual funding for individual international activities ranged from less than $1 million to over $560 million. The activities commanding the most resources were those that Calibre Systems associated with humanitarian and civic assistance, international agreements, and combined exercises (see Table 3.2).

The apparent contradictions in the findings by LMI and Calibre Systems highlight issues in the definition of Army international activities and their associated investment. LMI stated that security assistance programs involved the most resources of the three categories of the Army's international activities. This may be true, since between 2,000 and 3,000 Army military and civilian personnel are involved full time in security assistance activities carrying out transactions that total billions of dollars. But the level of resources that the Army must provide to such activities is really quite small, because fees paid by foreign governments on military sales cover almost all of the Army's costs. Another issue pertains to what type of activities the analyst should include in IA. Some Army activities may include international partners—or at a minimum have an international flavor—but can also be considered part of the Army's overall mission. For example, Calibre Systems construed civic and humanitarian operations to include both RESTORE HOPE in Somalia and similar operations in Rwanda. Although such activities further the national security strategy of engagement, they cannot be planned for in advance because

Table 3.2

Funds in the Army's 1995 Budget for International Activities

Activity	Funds ($ millions)
Security assistance[a]	
Foreign military sales	37
Military assistance program	1
Total security assistance	38
Political-military	
Foreign students at USMA	< 0.5
School of the Americas	2
Combined exercises	101
Overseas development training	28
Humanitarian and civic assistance	562[b]
International agreements	124
Status of forces negotiations	14
Foreign area officers	22
Total political-military	854[c]
Materiel-technology	
Foreign weapons evaluation	9
NATO cooperative R&D	2
Support of U.S.-FRG logistics	3
War reserve stocks for allies	4
Standardization and interoperability	11
Host nation support	73
Total materiel-technology	102
Overall total	994

[a]Includes only funds for unreimbursed costs.

[b]Includes funds for operations in Haiti ($188 million), Somalia ($344 million) and Rwanda ($28 million) for FY1994.

[c]Total without contingency costs for Haiti, Somalia, and Rwanda equals $294 million.

they are contingency operations and are outside of the Army's budget planning and management process.

The discrepancies between LMI's and Calibre Systems' findings point out the problems associated with identifying funds allotted to Army IA. Because the two studies used different definitions of types of resources and activities to include under the umbrella of international activities, they seem to have arrived at very different conclusions.[7] However, LMI's and Calibre Systems' findings with respect to resource levels for various categories are in much greater agreement once adjustments for these discrepancies are taken into account (see Table 3.3).[8]

Calibre Systems and LMI also agreed on other points. Like the LMI study, the Calibre study noted that resources for Army international activities come from multiple appropriations accounts (operations and maintenance, procurement, RDT&E, or the military construction accounts, although the preponderance is in the O&M account) and several program elements. Furthermore, within the Army's budget management system, Calibre Systems found that funding for IA is seen in many different MDEPs. These, in turn, are spread across multiple PEGs but make up only a small portion of the resources

Table 3.3

Comparison of Adjusted Resource Levels

Activity Category	LMI	Calibre Systems, Adjusted
Security assistance[a]	Largest	$150 million[b]
Political-military	> $300 million	$294 million[c]
Materiel-technology	Smallest	$102 million
Overall total	None given	$546 million

[a]Includes funds and resources not in the Army's budget.

[b]Does not include funds for military personnel.

[c]Excludes funds for operations in Haiti, Somalia, and Rwanda.

[7]LMI believed that security assistance activities commanded the greatest resources, while Calibre Systems' estimates allotted them the fewest resources of all three categories. LMI estimated that political-military activities received about $300 million annually, as opposed to Calibre Systems' estimate of over $800 million for 1995.

[8]Calibre estimated that a total of $150 million was devoted to security assistance programs in 1995 when some reimbursed costs were included. (This figure still excludes the reimbursed cost of active-duty military personnel.) Second, using Calibre's estimates but excluding funds for contingency operations in Haiti, Rwanda, and Somalia, results in a funding level for political-military activities of just under $300 million, which is close to LMI's estimate. And Calibre's estimate of resources devoted to materiel-technology activities—about $100 million—is indeed the lowest for the three categories.

managed by the respective PEGs. This situation and the fact that MDEPs associated with IA—such as the rationalization, standardization, and interoperability MDEP—are typically less than $50 million, mean that IA has a very low profile in the Army's budget.

RAND Efforts

We, too, attempted to identify the resources that the Army invests in its international activities and specifically those designed to enhance MFC. We did this in two steps. First, we tried to provide a context for efforts devoted to enhancing MFC within the Army's overall international activities by determining current levels of Army funding for IA. Then we attempted to identify those IA efforts that would most directly enhance MFC and the funding levels associated with them.

To accomplish the first step, we updated Calibre Systems' findings by determining the total resources that the Army has devoted to IA in its more recent budgets. But we made several adjustments to Calibre's methods as we updated their numbers. Because our research aimed to find ways to help the Army better manage its international activities that enhance MFC, we focused on funds that the Army can manage through its budget planning process. Since the Army does not pay for the costs associated with FMS or other security assistance programs out of its budgeted funds, we considered resources devoted to those programs outside of the Army's control and did not try to quantify them. Similarly, since the Army does not include funds for unanticipated contingency operations in its budget and cannot control the occurrence of such operations, we did not include resources for them as part of the overall level of resources available for Army international activities. Finally, although Calibre did not include funds devoted to the support of NATO operations in its estimate of resources devoted to IA, we felt that such activities and other similar activities could reasonably be included under the umbrella of Army international activities.

Using the criteria outlined above, we determined that funding levels for Army international activities in FY1999 were slightly less than $500 million (see Table 3.4). This would correspond to an adjusted

Table 3.4

Funds for International Activities in the Army's Budget

Activity	Funds ($ millions) 1995[a]	1999
Political-military		
School of the Americas	2	2
Chairman, JCS exercises	101	43
Overseas development training	28	N/A
Humanitarian and civic assistance[b]	2	5
International agreements	124	157
Status of forces negotiations	14	N/A
Foreign area officers[c]	22	30
NATO operations	242	183
Miscellaneous support to other nations	12	39
Materiel-technology		
Foreign weapons evaluation	9	13
NATO cooperative R&D	2	4
Support of U.S.-FRG logistics	3	N/A
War reserve stocks for allies	4	N/A
Standardization and interoperability	11	8
Host nation support	73	N/A
Overall total	650	484

[a]Based on adjusted Calibre Systems findings.

[b]Excludes funds for contingency operations in Haiti, Rwanda, and Somalia. Dollar amounts rounded.

[c]Includes funds for the Marshall Center.

N/A: Could not be determined from budget data available to RAND.

Calibre estimate of funding for 1995 equal to $650 million.[9] Although this lower funding level might represent a real decrease in resources devoted to IA, a more likely explanation is a change in Army

---

[9]Calibre Systems' estimated level of almost $1 billion devoted to Army international activities in 1995 included almost $40 million for security assistance programs and $560 million for contingency operations. In addition, the total excluded $254 million for support of NATO operations and similar support to other nations. After deducting funds for security assistance and contingency operations and adding funds for NATO operations support and similar activities, the adjusted Calibre Systems total for 1995 would be $650 million.

accounting and reporting practices.[10] In any case, out of a total Army budget for 1999 of $69 billion, we were able to identify approximately $0.5 billion earmarked for Army international activities.

Our second step was to determine how much of this approximately half-billion dollars was dedicated to improving the Army's ability to operate in multinational coalitions. Doing so depends, to some extent, on a subjective identification of those international activities that directly enhance MFC. Although all or almost all Army international activities might provide some indirect means for enhancing MFC, many fewer activities have a direct impact. Table 3.5 lists those forums that DUSA–IA has identified explicitly as a means to enhance MFC.

Even though it is difficult to isolate resources for even this small subset of IA, we determined that funding for MFC-related activities totaled slightly over $300 million for FY1999 (see Table 3.6). This figure, which represents about 60 percent of all Army international activities funds for 1999, is somewhat misleading, however. For ex-

Table 3.5

International Forums for Enhancing MFC, Identified by DUSA-IA/MFC

Exchange programs
NATO working groups
NATO standardization groups
ABCA standardization groups
Senior national representative visits
Bilateral staff talks
International cooperative opportunities
Combined exercises
International schools
International conferences

[10]We were unable to find any funds dedicated in 1999 to the host nation support activities ($73 million) or reserve overseas training ($28 million) identified by Calibre Systems for 1995. There was also a decrease in funding for joint and combined exercises of almost $60 million, which was due primarily to the absence of $49 million in funds for exercises programmed by Southern Command in Central America. The most likely explanation is that these funds are imbedded within other programs and cannot be isolated from data available to us.

ample, approximately $220 million of these funds were earmarked for the U.S. contribution to NATO's operating budget. These funds help pay for the Partnership for Peace program, but they also are used to help defray the operational costs for a wide array of NATO activities and programs such as the NATO AWACS, various multinational headquarters within NATO, and NATO communications.[11] It is impossible to determine from standard budget data how much of the funds allotted to NATO operations actually go to those programs that DUSA-IA has determined enhance MFC. Similar situations exist with respect to the funds for the other activities listed in Table 3.6.[12] On the other hand, funding for standardization programs and the Marshall Center are rather well documented, but together they represent a total of less than $40 million.

There were some MFC-related activities, such as visits by senior national representatives and international conferences, for which we could not identify any funding set aside within the Army's budget. We can only assume that the resources needed for these activities are so small that they are subsumed in the funding lines for larger programs.

Finally, there are undoubtedly funds devoted to some activities that contribute to enhanced capabilities in MFC that are not included in DUSA-IA's list of relevant forums (see Table 3.5). The foreign area officer (FAO) program is an example of such an activity. Foreign area officers act as liaisons, attachés, and security officers in foreign countries and with foreign militaries. The Army currently has about 800 FAOs and budgets several million annually for training them.

By our estimation, funding for MFC totaled about $310 million in FY1999, which represents about 60 percent of the total for Army

---

[11]These funds are not used to pay for the personnel costs of active-duty military personnel assigned to NATO organizations.

[12]For instance, funds for the Chairman of the Joint Chiefs of Staff exercise program—some $43 million in FY1999—are primarily used to support exercises overseas (90 percent, according to Army budget documents for FY2000). But RAND could not determine which of those exercises, and hence how much of the related funds, include forces from countries other than the United States.

Table 3.6

Funding of MFC Activities in Two Budget Structures

President's Budget		Army Internal Budget		
Activity	Funding Line	Type of funds	MDEP	Description
Exchange Programs ($220 million)				
Exchange programs	Misc. support to other nations, other support	O&M	XISQ	International nonmilitary assistance
	Support of NATO operations	O&M	XISH	U.S. contribution to NATO operations
Standardization and Interoperability ($10 million)				
NATO working groups & ABCA	Misc. support to other nations, tech transfer	O&M	VRSI	Standardization and interoperability
	Programwide activities	RDT&E		
NATO standardization	Support of NATO operations, standardization program	O&M		
Int'l cooperative R&D	Int'l cooperative R&D	RDT&E		
Combined Exercises ($50 million)				
Combined exercises	Operating forces readiness, joint & combined exercises	O&M	VJCS	JCS-directed exercises
Assistance to developing countries for exercises	Misc. support to other nations, humanitarian and civic assistance & DCCEP	O&M	JDHB	Developing country combined exercises program
The Marshall Center ($30 million)				
Marshall Center	Misc. support to other nations, Marshall Center	O&M	TAMC	The Marshall Center
	Operating forces, BASOPS	O&M		
	Operating forces, RPMA	O&M		
	Unified Commands	O&M		
Total funding ($310 million)				

NOTE: All dollar amounts are rounded to the nearest $10 million.

international activities.[13] In theory, then, if the Army decided to place a higher priority on MFC-related activities within the total funds allotted to Army international activities, there would be sufficient leeway to allow for shifting of resources.

There would be difficulty in actually doing this, however. The first reason, hinted at above, is that it is difficult to tie specific activities to exclusive funding streams or line items. Second, funding for many activities is included within larger programs and cannot be isolated easily. In short, funding streams as currently described in budget databases cannot be tied directly to programs, and vice versa. Second, many activities are not under direct control of the Army. For example, the schedule for exercises is established by the various CINCs, and the agenda for NATO operations is not set by the Army.

Given these constraints, the Army's internal budget management structure gives DUSA-IA only limited ability to act as a proponent for resourcing MFC-related activities. Calibre Systems identified 48 MDEPs that included funds related to Army international activities. Using a much narrower definition of activities that directly enhance MFC, we winnowed the number of MDEPs that include funds for MFC-related activities down to seven. Of these, DUSA-IA/MFC oversees only four (see Table 3.7), the largest of which is for NATO operations, which may include a preponderance of funds that are not allotted to activities that enhance MFC. On the other hand, it could be argued that joint and combined exercises are one of the biggest enhancers of MFC and thus are directly related to DUSA-IA's mission. DUSA-IA is not the proponent, however, for the MDEP containing funds for such activities.

Another aspect of the MDEPs currently under DUSA-IA/MFC oversight is that they are almost exclusively focused on NATO, Europe (the Marshall Center), or other established and developed allies. The TFAO and JVCS MDEPs have a much broader purview, encompassing over 100 countries in the former case and all major commands and theaters in the latter. However, other offices in the Army are proponents of these MDEPs. As a result, the current budgeting and

[13]If funds for support of NATO operations are excluded because they are used to support many activities that may not be related directly to enhancing MFC, then the total is reduced to about $100 million.

Table 3.7

Army Funds for International Activities Under DUSA-IA/MFC Oversight

Activity	MDEP	Funds in 1999 ($ millions)	DUSA-IA/MFC Oversight
The Marshall Center	TAMC	30	30
Combined exercises	VJCS	40	
Developing country combined exercise program	JDHB	*	
Foreign area officers	TFAO	*	
NATO operations	XISH	180	180
Miscellaneous support to other nations	XISQ	40	40
Standardization and interoperability	VRSI	10	10
Overall total		310	260

*Less than $5 million.

programming structure focuses most of DUSA-IA's attention on areas where the United States already has a close working relationship with its allies.

Another problem with the current budget structure is that data available to support it are not all that useful for establishing priorities among MFC activities. This is because there is little information on the geographic distribution of funds or the marginal cost per individual episode of an activity. That is, spending on MFC is not broken out by country, nor is the cost per individual conference or exercise specified.

Some examples may serve to illustrate these points. The first is CJCS exercises. The annual number of such exercises involving the Army has grown from 50 to 80 since 1990, with 90 percent occurring overseas. The range of countries involved has also increased to include central and east European countries and Middle Eastern countries such as Kuwait. First, DUSA-IA/MFC is not the proponent of these activities or their funding. Second, although it is possible to break out the distribution of funds by command, it is not possible, using the Army's budget database, to do so by country. Finally, budget data do not include the number of exercises in a given year or the countries involved. Thus, the cost of individual exercises or of country participation in an exercise cannot be determined. Without this information, the cost of adding a country to an exercise is hard to establish, as is the cost of running an additional exercise.

Another example is the International Military Education and Training (IMET) program, the funding of which is determined with little if any input from DUSA-IA. The IMET program provides education and training for foreign military and related civilian defense personnel both in U.S.-based Army schools and by mobile training teams in overseas facilities. The request for the annual allocation of spaces at Army schools or mobile training teams comes from the CINCs through the security assistance officers (SAO). TRADOC and DCSLOG coordinate assignments of spaces at the Army schools within CONUS, and the SAOs arrange in-country training. Thus, the SAOs and TRADOC (through its security assistance activities) play major roles in managing the process, while Army component commands and DUSA-IA do not. Required annual funding levels are formulated by the CINCs, and funds for the IMET program are appropriated separately by Congress.[14] Thus, DUSA-IA has little or no role in either the planning or resource allocation of this important international program.

The FMS program provides a final example of a program that has important implications for the performance of MFC but in whose management DUSA-IA has very little say. It is easy to see how sales of U.S. military equipment to likely coalition partners have the potential to improve the ability of the partners to fight together on the battlefields. But both the planning for the FMS program and the funds to administer those sales come from outside of the Army. A particular sale is initiated when a foreign country defines a materiel requirement and sends the U.S. Army Security Assistance Command a letter of request. Although Army personnel, both civilian and military, will implement the request, funds to cover the administrative costs come from the 3 percent surcharge assessed on each sale. Thus, although implementation of the FMS program imposes a cost on the Army, those costs cannot be measured in terms of budgetary resources and so cannot be easily compared with the costs of other activities.

---

[14]Funds for IMET and several other security assistance programs are included with the annual appropriation for foreign operations. Total funding appropriated for IMET in FY1999 was $50 million to serve more than 8,500 students from over 120 countries.

CONCERNS ABOUT ARMY RESOURCES FOR IA

Our review of available data and previous studies of Army international activities have identified several concerns about the Army's management of resources for IA. The first problem is that it is very difficult to determine the level of funding for IA, both in the aggregate and for individual programs. At least two other studies have highlighted this problem. The LMI report noted that the total personnel and resources involved in Army international activities were unknown. Likewise, the Calibre Systems effort concluded that the value of Army international activities could not be determined as a whole, nor could the individual programs be assessed for their value to the Army's plan for international activities.

Another problem is that the current management structure makes it difficult for DUSA-IA/MFC to be an effective proponent of some programs that enhance MFC because they are outside of DUSA-IA's purview. These programs include combined exercises and security assistance activities such as IMET. The former activities are planned by the CINCs but funded in the Army's budget. The latter are also planned, at least in part, by the CINCs, but they are funded outside of the Army's budget. These diverse planning and budgeting schemes make it difficult for DUSA-IA to coordinate and integrate all of the Army's international activities.

Finally, although budget data can provide some idea of the regional distribution of spending on Army international activities, they provide only a rough picture at best. Lack of information on funding or activities by individual country makes it difficult for the Army to know whether activities with some countries are under- or overemphasized.

IMPROVING ARMY MANAGEMENT OF IA RESOURCES

One of the things required for effective management of the Army's resources for IA is better communications between DUSA-IA and the offices that carry out the activities. Our main concerns about the current management of IA resources are that it is difficult to identify resources devoted to IA and that DUSA-IA/MFC does not have the ability to monitor or influence the allotment of funds for all relevant Army international activities programs. Thus, DUSA-IA needs better tools if it is going to be an effective proponent for activities that enhance MFC.

First, the Army needs to create a vehicle outside of the budget structure for collecting data on resources devoted to Army international activities. Funds for those activities are currently provided both off budget and on budget. Within the budget, they are scattered among many programs and MDEPs. Even if a way could be found to concentrate the funding streams in one place within the Army's budget, it would still not capture the significant resources provided for Army international activities from outside the Army, such as those for FMS. Thus, a separate mechanism is needed that will provide a complete funding picture for DUSA-IA, including the distribution of resources by activity, region, and country.

It is possible that the planned Theater Engagement Plan Management Information System (TEPMIS), or one similar to it, can provide such a mechanism. This on-line database, based on the European Command's Theater Security Planning System, is designed to provide a planning tool for the CINCs as they prepare their theater engagement plans.[15] TEPMIS will ultimately include a record of engagement activities, to include exercises, training, and security programs and other contacts, for all of the regional commands. As currently conceived, TEPMIS will include a two-year historical record as well as plans and requirements for the subsequent six years. Although the ultimate format for TEPMIS has not yet been established, some plans call for it to include funding requirements for the activities specified in the database. If such plans and requirements are indeed included, it will give planners the ability to search the database by activity, country, or region.[16]

---

[15]Every year, each of the CINCs draws up a theater engagement plan (TEP) strategic concept that outlines how he plans to carry out the strategy of engagement in his area of responsibility. The CINC lays out in detail the specific activities that he proposes to use to further his strategy in the activity annex to the TEP. TEPMIS is designed to support the creation and transmission of these activity annexes within each command and to the relevant planning and supervisory personnel.

[16]Some analysts have raised the concern that it might not be in the Army's interests to share preliminary budget data with the other services and OSD early in the budgeting cycle. One possible alternative or interim solution would be for the activity annexes that include budgetary data to be transmitted directly to DUSA-IA and other relevant

The TEPMIS system, or one similar to it, should provide consolidated data on Army international activities of interest to the CINCs. This would cover most of the activities that contribute significantly to enhanced MFC capabilities, such as exercises, officer exchanges, and the IMET program. Resources devoted to activities that may not be included in TEPMIS, such as those for cooperative R&D, are easily identified within the Army's budget. Combining such data with that from TEPMIS should provide a relatively complete picture of the distribution of resources for Army international activities, and go a long way toward correcting one of the current problems.

To address the second concern, that of finding more effective ways for DUSA-IA to influence the allocation of resources among the Army's international activities, there are several alternative structures available to the Army for managing the execution of its AIA programs. One option would be to have DUSA-IA play the role of advocate for MFC-related activities both within and outside the Army, but leave management of those programs decentralized. A second approach would give DUSA-IA the additional responsibility of establishing and disseminating policy guidance on how the various program managers should establish priorities for the use of their funding for international activities. In turn, the implementing offices would send information back to DUSA-IA on program implementation and budget needs. A third approach would be to put DUSA-IA in charge of both policy formulation and program implementation for all Army international activities.

The current structure is theoretically closest to the second approach outlined above. But reality does not follow the theory, and the current structure is in practice much closer to the first structure than the second. DUSA-IA is responsible for the preparation of the Army International Activities Plan (AIAP), which is supposed to provide guidance for establishing priorities among the Army's international activities. In its report, LMI also envisioned that various offices responsible for carrying out AIA could integrate the AIAP into their planning and budgeting. However, DUSA-IA, which is supposed to produce the AIAP annually, as of 2000 has not yet published a final (nondraft) plan. Its early draft plan was more along the lines of a

---

offices in the Army but to expunge budgetary data from those annexes submitted to TEPMIS.

"catch-all" document that includes little guidance to prioritize Army efforts in any systematic way. It is our understanding that the later draft plan had more elements in the guidance realm. In any event, DUSA-IA has not been providing clear and definitive guidance to the field. Second, as was discussed earlier, DUSA-IA has not been receiving the broad feedback from the field that it needs to set informed policy. Nor does DUSA-IA have any role at all in planning some activities such as those associated with FAOs or CJCS exercises. As a consequence, DUSA-IA's role has become limited to that of an advocate for AIA with much less of a role in determining policy guidance or establishing priorities.

The final alternative management structure we considered would give DUSA-IA a greatly increased role in the management structure of all AIA. Although this might seem to be the most efficient structure and would provide a single focal point within the Army for all international activities, it would also have several drawbacks. The first is that it would be very disruptive of current programs that have evolved within the current structure, some over several decades. During that time many working relationships and practices have been established that it might prove disadvantageous to disrupt. Second, centralizing management in DUSA-IA would remove responsibility from the field, and this, given the far-reaching and highly diversified portfolio of Army international activities, would undoubtedly lead in some cases to poor decisions being handed down from the top. In part for these reasons, LMI recommended against such a centralized structure in its 1996 report, and we would make the same recommendation now.

Rather, we recommend a strengthening of the two-way relationship between DUSA-IA and the field offices that is necessary to implement the second option. Although we are not arguing that DUSA-IA should control funding for all IA, we are arguing that it should have visibility of and influence over the allocation of funding for IA, particularly activities that directly enhance MFC.

One mechanism would be to expand the number of MDEPs for which DUSA-IA has some primary advocacy. DUSA-IA/MFC now has oversight of the allocation of resources for only four MDEPs. However, it can be argued that DUSA-IA should have input into the planning and resource allocation of several other MDEPs—including those relevant to the FAO program and those related to exercises.

DUSA-IA could also use the AIAP as a vehicle for codifying its policy and disseminating it to the field. Those offices could then, in turn, use the AIAP and DUSA-IA's guidance to set budget priorities for their own activities. LMI recommended such a process in which DUSA-IA would first disseminate integrated planning guidance to the various parts of the Army. In response, the various implementing commands would prepare and submit plans to DUSA-IA for integration into the AIAP. This, in turn, would be used as a blueprint for budgeting Army international activities. That is, the functional proponents would integrate Army international activities into their planning and budgeting documents. DUSA-IA would then be responsible for monitoring the activities and assessing program effectiveness throughout the process for input to the next version of the AIAP. In this way, DUSA-IA would provide guidance on both the activities to be undertaken and allocation of resources.

For this approach to work, two changes would have to occur. First, DUSA-IA must issue clear and specific policy guidance on a timely and frequent basis. Second, the implementing offices and commands would have to provide, on a correspondingly frequent basis, information to DUSA-IA on program resource requirements and program effectiveness. The process will not work without both inputs, as is demonstrated by the current state of affairs.

It is evident that guiding, monitoring, and integrating all Army international activities is a big task, and perhaps DUSA-IA is already on its way to implementing this recommendation. Currently, however, DUSA-IA has neither the means to assess priorities nor a process in place to monitor and influence the planning and budgeting for the wide range of activities that enhance MFC. Thus, DUSA-IA first needs to take steps to collect all the pertinent resource data on an annual basis. Having done so, the next step will be to set up a process to monitor the implementation of its guidance in the budgeting and planning documents of the various commands charged with executing the Army's international activities. Once DUSA-IA has completed these two tasks it will be in a better position to evaluate the costs and benefits of the various activities that enhance the Army's performance in MFC.

Chapter Four

AN INTEGRATED SYSTEM OF PLANNING MFC EFFORTS

For DUSA-IA to be effective in its role as the main proponent of the Army's MFC-related activities, it needs to have an overall blueprint or vision of what it seeks to accomplish. Without such a blueprint, even a doubling or tripling of funding for Army international activities does not guarantee a commensurate increase in the effectiveness of the Army's MFC efforts, nor is there any guarantee that the funds are being spent wisely. This chapter addresses the issue of an overall blueprint that would guide the Army's MFC planning process.

Currently, the institutional Army does not have an explicit long-term plan for its MFC efforts. Army personnel play a supporting and implementing role in carrying out the MFC activities they are asked to do by the Joint Staff and the CINCs, with the funds usually coming out of the Army's budget. The absence of an Army-based long-term MFC plan is a shortcoming for a number of reasons. First, the various Army activities do not necessarily add up to a unified whole from the perspective of the institutional Army. An overall plan and guidelines are needed to unify and give more coherence to the various Army MFC activities. Second, there is a large middle ground between the guidelines for MFC efforts established by OSD and the specific TEPs of the CINCs. Guidelines based on the institutional Army's long-term views on MFC activities would go further than the general OSD guidance but be more general and long term in outlook than the TEPs. Third, Army component commands have an input in the TEP process, but currently they have in mind few, if any, institutional Army goals for MFC. The institutional Army's guidelines for MFC would be important input for planning discussions of MFC efforts at the joint level or with CINCs. Individual Army commands and agencies currently have their own ideas and standards for advancing MFC, but what is lacking is an overall Army MFC blueprint that would integrate the various Army efforts. Without such a blueprint, there is a real danger of inefficient outcomes and duplication of efforts or, at a minimum, a lack of synergy between the various Army activities designed to enhance MFC.

All three services have launched efforts recently to put together service-based MFC prioritization plans. The Center for Naval Analyses provided analytical assistance on such a plan for the Navy, while DFI International conducted similar work for the Air Force.[1] DUSA-IA has defined the dimensions of what an Army international activities plan would need to have. Although the existing efforts emphasize the role of allies in coalition operations with the United States and attempt to establish the importance of specific states to the given service, none of them deals adequately with the problem of forecasting the most likely coalition partners and targeting specific efforts at those states. It is not a given that allies and coalition partners will necessarily be the same. In addition, none of the existing efforts that we are aware of goes to the next stage of identifying specific steps to take in order to increase compatibility.

What we propose here is a planning process designed to help the Army establish and implement overall MFC guidelines. The process consists of four steps:

1. Prioritize the Army's MFC efforts by ranking countries on the basis of the most efficient targeting of scarce resources.
2. Ascertain the compatibility deficiencies of the individual partner countries, taking into account the type of operation in which the given country's forces are likely to participate with Army forces.
3. Link appropriate Army MFC policies to address the identified compatibility deficiencies.

---

[1]For the Navy, see Marvin A. Pokrant, Jonathan D. Geithner, Catherine K. Lea, Patrick Roth, *Framework for Allocating OPNAV's International Engagement Activity*, Alexandria, VA: Center for Naval Analyses, CRM 98-106, October 1998. For the Air Force, see *Assured Access: Constraints and Prospects in the Next Century*, briefing, Washington, D.C.: DFI International, July 1998. The closest recent effort along these lines undertaken by the Army (that we are aware of) is Gordon Yim, *America's Army: Moving into the 21st Century, Army International Activities*, briefing, DAMO-SSP, undated.

4. Determine cost-effectiveness in the allocation of resources to activities designed to further MFC.

Step 1 is global in nature and requires consideration at the level of the Department of the Army. Prioritization stemming from this step would provide guidelines to Army component commands in their input to the TEP process. The other steps can be carried out by the Army component commands in cooperation with DUSA-IA. In addition, at the aggregate level, DUSA-IA needs to be aware of the pattern, effectiveness, and cost of efforts, by country.

STEP 1: PRIORITIZING COUNTRIES BY PROPENSITY TO BE COALITION PARTNERS

To design a plan for prioritizing the Army's partners, the best way to proceed is to start with a clear idea of the goal. We should therefore ask: What kind of "end state" to its MFC efforts would the Army wish to see from the perspective of its Title 10 functions of preparing the force? Ideally, and in the most general sense, the Army would wish to be able to build an effective fighting (or peace operations) force rapidly with the coalition partners participating in a deployment. In many recent operations the United States has had the luxury of lengthy time to prepare for actual operations, but that luxury may not exist in future operations. The ability to build an effective multinational force entails knowing in advance the appropriate level of cooperation between Army units and units from partner countries. This in turn entails the knowledge ahead of time of a particular country's likelihood of being a coalition partner and its importance in a coalition operation. Furthermore, it entails the knowledge of the expected partner units' level of compatibility with Army units, so as to assign an appropriate role to the partner forces.

With the above "end state" in mind, an ideally efficient Army MFC policy would target just the right countries (the ones that would be substantially and substantively involved in an actual coalition operation) and just in the appropriate areas (ones useful to fulfillment of the role assigned to the units in an actual coalition operation). For a variety of reasons, an ideally efficient policy is not achievable. First, it is impossible to predict with certainty which countries will participate in a coalition operation with the United States during the next decade for the simple reason that it is impossible to predict fully the coalition operations in which Army units will participate. The list of countries likely to participate can be narrowed down, but there will always be a continuum of likelihood of participation, and outliers will sneak in at least occasionally (for example, Syria in the Gulf War). Predictions become more difficult and more prone to error the longer the time frame examined. The same applies to the likely role in a coalition. For political reasons, Army units may have to be integrated with some partners to an extent that the Army would prefer to avoid under "normal" circumstances. Limits on capability will constrain the level of integration possible, but the limits need to be seen as a range on a continuum rather than a specific point.

In addition, the Army will cooperate in and spend resources on MFC activities with countries that are unlikely to be expeditionary coalition partners while fulfilling policies of DoD, the Joint Staff, or CINCs (policies often designed more with political than military objectives in mind). From the perspective of Army MFC goals such activities may be "inefficient," in that they do not represent the core of the service's multinational force compatibility strategy, but because they serve the larger U.S. political-military objectives of regional deterrence or engagement, the Army will need to carry them out. In this vein, it is also important to consider certain national-level commitments as well as recognized threats that entail Army MFC activities with some countries. For example, the defense of Kuwait and South Korea necessitate preparation and MFC activities by the Army in order to deter conflict and to ensure effective cooperation with the Kuwaiti and South Korean armed forces if a need arose to make good on U.S. commitments. The national-level commitments allow no leeway or ambiguity in Army MFC activities.

Thus, the goal to aim for is a prioritization that comes closest to the ideal set of policies, both in the realm of country priorities and the types of efforts undertaken, keeping in mind that the prioritization is subject to periodic change and regular reevaluation. In addition, even in areas where the Army has little say regarding country or policy choices, Army planners need to be aware of the larger institutional priorities, so as to take advantage of the synergies and think in terms of the totality of MFC-related efforts. Since, other than Kuwait and South Korea, there are few clear and recognized threats that would lead to prioritization choices, the first and most difficult question is how to figure out ahead of time the likelihood of a given country's participation in a coalition with the United States. The next section proposes a methodology for such a prioritization.

Rationale for Participation in Coalition Operations with the United States

The central problem in anticipating the makeup of a coalition centers on the nature of coalition formation. At the heart of the matter is the collective action problem, or the paradox that even actions clearly in the collective interest will not be undertaken because individual-level incentives act to curtail participation. The problem arises in conditions of "nonexcludable public goods," where an action's benefits cannot be excluded from applying to all members, even if only a few have contributed to creating the benefits. Thus, each actor has an incentive to "free ride," i.e., minimize one's own input and expect another member to make the effort and pay the costs of the collective action. The only way to encourage participation and to reduce "free riding" is through "privatizing" the benefits or, in other words, restricting the access to benefits only to those who contributed to their production. The collective action problem has been useful in explaining intra-alliance behavior, namely the tendency of the largest and most affluent members of an alliance to contribute proportionately more to the collective defense provided by the alliance. Of course, there are important differences between alliances and coalitions. Alliances have formal arrangements for cooperation that serve peacetime purposes of deterrence and enhancement of the conduct of potential warfighting. Coalitions, on the other hand, are ad hoc and temporary arrangements for common action between two or more states, formed for a specific purpose and usually dissolved once the purpose is achieved. However, both the collective action problem and the questions of what makes states most likely to align in the first place underlie the efficient organization of effort within coalitions and alliances.

The collective action problem illustrates the core problem of finding meaningful coalition partners in light of the hegemonic[2] nature of

[2]We use the term "hegemon" in a value-free fashion to refer to the particular state that has a preponderant degree of influence in the international state system.

U.S. power in the contemporary state system. Since 1990 the world has been largely unipolar, with a preponderance of U.S. military might, a condition likely to persist for at least the near and middle term. Barring the unlikely emergence soon of a "peer competitor," the situation may persist for the long term (beyond 2015). Given its enormous military strength relative to any other single state, if the United States makes it clear that it has an interest in a specific operation, other states have incentives to "free ride" or "easy ride" and let the United States conduct an operation largely on its own, or with whatever other states will participate. This is true even in conditions whereby other states also stand to gain from the benefits of the operation. Moreover, it is particularly true of the prevalent form of contemporary coalition operations—primarily peace operations—in which the United States is engaged. In these operations, the security benefits are especially diffuse, contributing to a difficult-to-define overall global or regional stability rather than alleviating clear threats to sovereignty of any states (other than, perhaps, those where the operation takes place).

Overcoming the Collective Action Problem

The crucial question to ask, then, is which states, and under what conditions, are apt to be less prone to the collective action problem, assuming the continued U.S. hegemonic role. A primary rationale for a state to join a coalition operation is that it sees direct benefits to participation that would not otherwise be attainable. In other words, the benefits of participation are at least partially "privatized." Probably the clearest example of partially "privatizing" the benefits of coalition operations entails the practice of payment to states that contribute forces to UN peacekeeping operations. In the context of extremely poor countries, the revenue-generation function of coalition membership (and perhaps the training benefits of participation in such activities) is a nontrivial benefit and explains why some of the poorest countries in the world, such as Bangladesh or Ghana, consistently provide forces (up to a third of their peacetime force size) for such operations.

But there are other ways of "privatizing" the benefits, related most of all to the reasons why states choose to involve themselves in conflict and war. In a situation of ever-present competition for power among states, the hegemonic position of the United States, although a disincentive for some states to participate in U.S. coalition operations, also offers other states a specific rationale to join in. For that to happen, a state must see a link between participation in a coalition with the United States now and security ties with the United States in the future. If a state expects that its U.S. ties will be stronger as a result of such participation, and sees the maintenance and strengthening of those ties to be in its interest, then there is effectively a private good to be gained from contributing. Thus, the preponderance of U.S. power may make some states, particularly the ones that face potential regional threats or are otherwise uncertain as to their security, see an extended security relationship with the United States as a means of deterring their regional competitor(s) or threat(s).

Congruity of interests may sometimes be even more important than alliance membership. In the case of states that aspire to a U.S. security commitment but have not obtained it because of U.S. reluctance, performance as a reliable partner in coalition operations may be a way for such a state to demonstrate its importance and usefulness. In other words, a desire for a U.S. security commitment leads to conditions of dependency and concern about remaining in good stead with the United States, which may be sufficient to elicit participation in a coalition operation where the United States plays a major role. An example of this phenomenon is the participation of many Partnership for Peace (PfP) states in IFOR/SFOR. Another example is the backing of NATO's Kosovo operation by the PfP countries aspiring to NATO membership, such as Slovakia or Romania, which gave NATO flyover rights despite their having no bilateral problems with Serbia. Contrast that behavior with the outright refusal by Greece to allow combat aircraft to overfly its territory in support of the Kosovo mission.

In our effort to establish which states will overcome collective action problems to participate in multinational coalitions with the United States, one obvious indicator is past participation. However, with only a single decade of experience with the contemporary world situation, we must look for other proxy measures that will help us predict future proclivity to participate.[3] Common alliance membership with the United States and development of defense ties provide such an indicator, though only a partially useful one. There is an ongoing debate as to how often countries join in the fight when their allies engage in military operations. The most detailed empirical study of the issue concluded that allies are reliable about 75 percent of the time.[4] Earlier studies, which did not take into account specific treaty limitations of alliances, showed that allies appeared to join their partners in war about a quarter of the time.[5] In any event, although alliance membership alone is not a decisive determinant of propensity to join a coalition, allies do join in the fight at a rate that is better than one might find for randomly selected nonallied states. Among the attempts to ferret out why allies sometimes join a fight, the two most relevant for our purposes of attempting to predict coalition composition are the findings that newer, defensive alliances are generally more effective and that conflicts which invoke the ostensible cause of the alliance are more likely to draw in more alliance members.[6] This makes intuitive sense, supporting long-standing analysis defining alliance formation predominantly as a response to perceived or actual threat to the countries involved.[7] Thus, the age and purpose of a given alliance provide additional insight into the likelihood of its members taking part in a coalition operation. Finally, existing alliances bring with them ongoing joint exercises and mili-

---

[3]We recognize that the U.S. alliance system is a legacy of the Cold War and that relationships established during the Cold War may have a bearing on propensity to participate in a coalition operation alongside the United States. However, the nature of the international system and of coalition operations is fundamentally different since 1990, and the experiences and patterns of the past decade (rather than the past half-century) are most relevant to calculations and projections about the next decade.

[4]Brett Ashley Leeds, Andrew G. Long, and Sara McLaughlin Mitchell, "Reevaluating Alliance Reliability: Specific Threats, Specific Promises," *Journal of Conflict Resolution,* Vol. 44, No. 5, October 2000, pp. 686–699.

[5]Alastair Smith, "To Intervene or Not to Intervene," *Journal of Conflict Resolution,* Vol. 40, No. 1, March 1996, p. 17. Randolph M. Siverson and Joel King, "Attributes of National Alliance Membership and War Participation, 1815–1965," *American Journal of Political Science,* Vol. 24, No. 1, February 1980, found that 23 percent of alliance members joined in when an alliance partner was a belligerent.

[6]See Siverson and King; Chae-Han Kim, "Third-Party Participation in Wars," *Journal of Conflict Resolution,* Vol. 35, No. 4, December 1991, pp. 659–677.

[7]Stephen M. Walt, *The Origins of Alliances,* Ithaca, NY: Cornell University Press, 1987; Kenneth N. Waltz, *Theory of International Politics,* New York: McGraw-Hill, 1979.

tary planning. This bodes well for the effectiveness of military operations undertaken in concert by alliance members and thus argues for a capability, if not necessarily a proclivity, for operating effectively in concert. In fact, this may make it more likely that states will join a coalition together, reducing the transaction costs of forming a coalition by ensuring that they will not face as much difficulty coordinating their operations as they would with states that lacked a shared history of joint exercises and military planning.

Congruity of regional and global interests with those of the United States is another indicator of propensity to join it in a military coalition operation. Other than a few powerful states with supraregional security interests, states are most likely to become involved in operations regionally and generally close to their own borders. Smaller states seldom have more than regional threat perceptions. Within their region, however, they may see a range of actions as vital to the ensuring of their own territorial defense. The actions of Bahrain, Saudi Arabia, and the UAE to help reverse the 1990 Iraqi land grab in Kuwait provide an example of states participating in military operations to defend the territory of another from a common enemy in order to weaken that enemy before it can expand its reach. The Gulf War also illustrates the heavy regional presence of the coalition forces; while the coalition included forces from dozens of countries, the Middle Eastern countries were the most heavily represented. Thus, state views on security issues, both regional and global, can serve as indicators of the extent to which another state's interests are similar to those of the United States, and therefore of the likelihood that the state will see similar threats and wish to respond to them similarly.

Common ideology, in this case a democratic political system, provides another indicator of propensity to join the United States in a coalition operation. The idea is based on the "democratic peace" proposition, one of the most hotly debated theoretical issues in international relations theory. "Democratic peace" supporters generally cite the low rate of war between democracies and argue that democracies are therefore less prone to war with each other, due to cultural and normative reasons, institutional constraints, and/or some form of economic self-interest.[8] Critics of the proposition counter these arguments by noting that such institutional explanations would predict a lower rate of war for democracies overall, rather than just with each other, that the seeming lack of inter-democracy war is debatable (especially as many proponents selectively define both "democracy" and "war"), that a close study of cases of "near war" obviates the argument that democracies truly seek to avoid fighting each other, and that regardless of the proclivities of full-fledged democracies, democratizing states may actually be more, rather than less, prone to conflict.[9] Theorists also generally discount ideology as a significant factor in coalition formation in the context of systemic power/threat calculations.[10] Finally, even accepting the democratic peace proposition, it is another leap to claim that democracies are therefore more likely to participate in a coalition with the United States. That said, there is a good deal of at least correlational association between democracy and lack of conflict between those states,[11] there are increasingly persuasive causal

---

[8]Walt; Waltz; Carol R Ember, Melvin Ember, and Bruce Russett, "Peace Between Participatory Polities," *World Politics*, July 1992; and Michael W. Doyle, "Liberalism and World Politics," *American Political Science Review*, December 1986. See also Michael W. Doyle, "Kant, Liberal Legacies and Foreign Affairs," Bruce Russett, "The Fact of the Democratic Peace" and "Why Democratic Peace?" and John M. Owen, "How Liberalism Produces Democratic Peace," all in Michael E. Brown, Sean M. Lynn-Jones, and Steven E. Miller (eds.), *Debating the Democratic Peace*, Cambridge, MA: MIT Press, 1996.

[9]See Christopher Layne, "Kant or Cant: The Myth of the Democratic Peace," David E. Spiro, "The Insignificance of the Liberal Peace," Henry S. Farber and Joanne Gowa, "Polities and Peace," Ido Oren, "The Subjectivity of the 'Democratic' Peace: Changing U.S. Perceptions of Imperial Germany," and Edward D. Mansfield and Jack Snyder, "Democratization and the Danger of War," all in Brown, Lynn-Jones, and Miller (eds.), *Debating the Democratic Peace*. See also the additional "point and counterpoint" articles in that volume.

[10]Walt provides a useful overview of the literature on ideology and alliance formation in his discussion of the relevant hypotheses, pp. 33–40, 206–212. See also Michael W. Simon and Erik Gartzke, "Political System Similarity and the Choice of Allies: Do Democracies Flock Together or Do Opposites Attract?" *Journal of Conflict Resolution*, Vol. 40, No. 4, December, 1996, pp. 617–635; Ole R. Holsti, P. Terrence Hopmann, and John D. Sullivan, *Unity and Disintegration in International Alliances*, Lanham, MD: University Press of America, 1985 edition, pp. 61–85; and Glenn H. Snyder, *Alliance Politics*, Ithaca, NY: Cornell University Press, 1997, pp. 142–143.

[11]In over 100 empirical studies of the democratic peace theorem conducted during the past ten years, none has identified a "positive and statistically significant relationship between democratic dyads and militarized conflict." William R. Thompson,

explanations of the "democratic peace" theorem,[12] and the theorem is one of the few broad conclusions largely accepted as valid by mainstream political science.[13] And while ideology is difficult to defend as decisive in the selection of allies and partners, there is reason to believe that it may have some bearing, and some of the specific ways in which ideology can have an impact are particularly interesting. For instance, it has been noted that ideology does make a significant difference in the *form* an alliance will take. Specifically, states with similar ideologies are more likely to form military coalitions rather than, say, nonaggression pacts.[14] Furthermore, in explaining why the Cold War did witness a fair amount of alignment seemingly along ideological lines, it has been argued that the Cold War's bipolar structure, by increasing overall security, enabled the formation of at least some alliances based on nonsecurity factors such as ideology.[15]

The ideological element is important because in the contemporary unipolar state system, many of the states that are candidates for coalition formation with the United States are, like the United States itself, relatively secure when compared with the Cold War era. Furthermore, the coalitions in question are often formed for reasons not directly linked to many participants' security, i.e., peacekeeping and humanitarian operations. For these efforts, a uniquely security-based analysis may be insufficient. The finding, for instance, that alliances function best when their cause is invoked (noted above) may be moot for peace operations, which are not the type of operations that generally spur formal alliances.[16] Conversely, the propensity of ideology to help determine the shape of a coalition—

Richard Tucker, "A Tale of Democratic Peace Critiques," *Journal of Conflict Resolution*, Vol. 41, No. 3, June 1997, pp. 428–454.

[12]William Reed, "A Unified Statistical Model of Conflict Onset and Escalation," *American Journal of Political Science*, Vol. 44, No. 1, January 2000, pp. 84–93.

[13]James Lee Ray, "Does Democracy Cause Peace?" *Annual Review of Political Science*, Vol. 1, 1998, pp. 27–46.

[14]Holsti, Hopmann, and Sullivan, pp. 61, 64.

[15]See Walt, p. 38; Simon and Gartzke.

[16]Contemporary NATO especially faces this as an acute problem. Todd Sandler, "The Future Challenges of NATO: An Economic Viewpoint," *Defence and Peace Economics*, Vol. 8, 1997, pp. 319–353; Joseph Lepgold, "NATO's Post–Cold War Collective Action Problem," *International Security*, Vol. 23, No. 1, Summer 1998, pp. 78–106.

particularly to ensure that it is a commitment to cooperate militarily—and to have more of an impact when risk is low makes it a factor worthy of inclusion. It is, in fact, military coalitions formed by states *not* facing clear and direct threats that are at issue in many peace and humanitarian operations.

Finally, the hegemonic position of the United States in the international state system does not mean that it is the only "privileged" state in the international system. States that do not aspire to hegemony at the level of the international state system but consider themselves "great powers" or leaders on the world stage will act accordingly. French and British participation in the Gulf War may be an example of this phenomenon, as may Russia's continued willingness to participate in peace operations in the former Yugoslavia, despite its relegation to distinctly junior status. In areas where the United States is less involved, regional "hegemons" may seek to play a similar role. An example is Nigeria's peacekeeping efforts in Africa[17] or Australian efforts in the southwestern Pacific (Papua New Guinea, Solomon Islands, East Timor/Indonesia). While such states' interests are fundamentally regional, they might be enticed into coalition operations alongside the United States if, for instance, the private good of a leadership role can be offered. This would raise their profile in their region, which may be seen as translating into increased power and influence.

Using these conclusions and analyses to create a useful prioritization of states according to their propensity to join the United States in future ground-force coalition operations requires transforming them into a set of indicators. First, we recognized that some states could not, or are currently unwilling to, contribute to ground-force coalition operations at a level that would justify any significant expenditure of Army efforts on MFC activities with these countries from the perspective of the institutional Army. We eliminated from consideration those states with a particularly unsettled domestic situation, such as active civil war, since such countries have other—from their

---

[17]Carolyn M. Shaw and Julius O. Ohonvbere, "Hegemonic Participation in Peace-Keeping Operations: The Case of Nigeria and ECOMOG," *International Journal on World Peace*, Vol. 13, No. 2, June 1996, pp. 31–60.

perspective more urgent—tasks for their troops.[18] We also eliminated from consideration clear adversary states of the United States.[19] Then, calculating that a minimum of one battalion is necessary for a state's contribution to be meaningful to a coalition, and considering that a state would be unlikely to send over 25 percent of its ground troops abroad as part of such a multinational coalition, we selected for consideration only those states that had a force size of 5,000 ground troops in peacetime. We then divided the states in our sample into three categories, high, medium, and low, based on ground troop strength above that baseline, reflecting the ability to contribute a brigade, a division, or a larger force to the overall effort. The tool we put together is dynamic, and if any of the states we eliminated because of their internal problems or adversarial stance toward the United States were to change in status or in force strength, the framework can easily adapt to consider them for MFC efforts.

Second, we distinguished among countries by the level of technological sophistication of their armed forces. We used the measure of per-soldier spending on armed forces in the most recent year available as a proxy indicator for the level of overall weapons technology and coordination within a state's ground forces. We took a high ranking on this score (defined as equivalent to 50 percent or more of U.S. per-soldier spending) to mean that the forces of a given state had equipment sufficiently sophisticated to allow the United States to confidently use them in roles it would itself undertake, regardless of the intensity of the operation on a mission spectrum.[20]

---

[18]Only some states undergoing civil strife were excluded, with the main consideration for exclusion being that the conflict affected the functioning of the country in a fundamental fashion. Thus, countries such as Colombia, Liberia, Algeria, Afghanistan, Sudan, and Angola were excluded. Others, such as Turkey, Azerbaijan, and Uganda, were included because civil strife in these countries tends to be localized and has not crippled the functioning of the state in general. Generally, we considered an ongoing major UN peacekeeping operation in a given country as enough reason for exclusion from consideration. For a detailed listing of the countries excluded on the basis of civil strife, see Appendix A.

[19]At the time of completion of this report, the group of current "clear adversaries" (countries against which the United States has wide-ranging sanctions in place) included North Korea, Iraq, Iran, Libya, Yugoslavia, and Cuba.

[20]To be sure, an army's greater capital intensity may not translate into greater compatibility. This variable was not included as a proxy of compatibility; rather, it serves as an indicator for this military's degree of sophistication relative to the U.S.

Third, in an effort to consider what countries have the shared regional and global interests necessary to become contributors and which ones are susceptible to the privatization of additional goods that will help overcome the collective action problem endemic to coalition formation, we selected several indicators as proxies.

In addition to looking at the history of participation in main military coalition operations with the United States over the past ten years, we constructed a proxy indicator based on formal and informal security ties with the United States. In doing this, we paid particular attention to new alliance members or countries aspiring to alliance with the United States, where applicable. To provide further background on shared security perceptions and general worldviews that could be translated into military action, we traced the voting patterns in the UN General Assembly annually for each country since 1990. We used two types of voting actions: voting coincidence with U.S. votes in overall General Assembly actions (excluding consensus votes), and voting coincidence with U.S. votes in the General Assembly identified as "important" by the State Department. Numbering no more than a dozen per year, the "important" votes are generally on security issues that directly affect important U.S. interests, and the State Department lobbies other countries extensively to vote with the United States. Consistently not voting with the United States on such resolutions is an indication of probable differences in security outlooks. We judged the level of common ideology with the United States by measuring the level of democratization, on an annual basis since 1990, as reported by Freedom House in its annual "freedom ratings."

Altogether, we considered 108 countries that met the initial thresholds. We ranked each country on a standard scale by regional groupings. The regional basis is important because proximity elevates the sense of threat, and some countries may be important regional partners for the United States. The section below discusses the prioritization that emerged from the data. Detailed description

Army. There also may be cases in which high per-soldier spending does not imply greater sophistication—indeed, a country could be spending too much on equipment, or not coupling capital investment with adequate training to operate such equipment. Army planners need to be mindful of these cases, but the data demonstrate that, as a general rule, greater per-soldier spending translates to greater sophistication.

of the methodology, ranking criteria, data sources, and a full country-by-country presentation of data is included in Appendix A.

Combining these indicators presented additional challenges. All of the factors examined in this assessment are important but they are not weighted equally when putting together the resulting choices. Although high scores across the board would represent an ideal partner, probably the most important single consideration is past participation. Consistent participation in coalition operations with the United States over the past decade is not a proxy indicator; it shows proven willingness to contribute forces. Unless there is a major internal political shift in a country, its pattern of action over the next decade is likely to be similar to the pattern for the previous decade. Similarly, military ties and alignment, an indicator of proclivity to conduct military exercises and strength of relations with the United States, is a real measure of willingness to engage militarily with the United States and forms an important factor in assessing a given state's propensity to join the United States in a coalition. Ground force strength shows the level at which a given state can theoretically contribute to a coalition. In some cases, only small contributions might be required, but, all other things being equal, a major potential contributor deserves greater MFC efforts than a minor one.[21] Per-soldier spending is to a large extent an indicator of a given military's sophistication and propensity to contribute substantively to missions across the spectrum of conflict. The two sets of UN voting records and the democracy ratings constitute proxy indicators of likely future behavior. They frame the general levels of support to the United States in coalition operations that might be expected in the future. Recent trends in the voting and democracy indicators also foreshadow potential changes in past behavior in the propensity to join the United States in a coalition operation.

---

[21]Some small countries may wish to role-specialize in the provision of specific functional assets/skills to a coalition operation, such as air defense or C4I. Significant niche contributions should certainly be considered and targeted, and the Military Compatibility Assessment Tool (MCAT) discussed below allows Army planners to adjust the weight placed on a country's capability in different functional areas, such as maneuver, C4I, and logistics. At the same time, however, militaries usually seek to maintain a balance between the level of development of different functions, so that a country able to contribute a C4I battalion is in all likelihood able to provide other similarly sized units.

Some might argue that weighting past participation and current alignment as the two most important factors biases the results by raising the importance of Europe and of NATO, since four of the seven major previous coalition operations we considered were NATO actions and took place in Europe.[22] While the bias and the consequent emphasis on Europe and the European allies does influence the analysis, it is also true that Europe constitutes an area of declared U.S. strategic interest, and the United States has alliance and other security commitments to many countries in Europe. Indeed, it is no accident that most of the operations took place on the European continent.

Results of Prioritization

We classified the countries that scored high in the prioritization process as either worldwide or regional partners. The countries listed in Table 4.1 scored the highest on the set of variables measuring the propensity to offer a minimum effective contribution to future coalition operations involving the United States.

Before proceeding further, it is important to emphasize that the methodology presented here for prioritizing countries for MFC purposes is *not meant to be and should not be* used in a mechanistic fashion. It is a tool that provides a shortcut for planners to identify the most likely coalition partners. It enables planners to portray results on a global basis but does so at the price of some abstraction and a loss of consideration of specific country information. The methodology is not a substitute for common sense and more detailed country-specific knowledge. As mentioned earlier, the methodology does not take into account the national-level commitments and specific recognized threats because the MFC activities stemming from such considerations are a given for the Army. For example, Kuwait or Colombia did not make the priority list (since the armed forces of these countries are unlikely to be sent on an expeditionary coalition operation), but Army planners must pay close

[22]The four NATO operations in Europe were Deliberate Force (Bosnia-Herzegovina), Joint Endeavor (IFOR, Bosnia-Herzegovina), Allied Force (Kosovo), and Joint Guardian (KFOR, Kosovo). The non-NATO and non-European operations were Desert Storm (Gulf War), Restore Hope (Somalia), and Uphold Democracy (Haiti).

Table 4.1

Summary of the Prioritization Results

Worldwide Partners	
First tier	• United Kingdom • France • Germany
Second tier	• Canada • Australia • Belgium • Netherlands • Italy • Spain
Regional Partners	
Europe	• Other NATO allies • Select PfP countries (e.g., Sweden, Romania)
Middle East	• Turkey • Israel • Egypt • Jordan • Morocco • Saudi Arabia
Africa	• Botswana • Nigeria • South Africa • Some ACRI states (e.g., Senegal, Ghana)
Central and South Asia	• No readily identifiable long-term partner
East Asia and Pacific	• Japan • South Korea • Singapore • Thailand
North and Central America	• El Salvador • Honduras • Guatemala • Nicaragua
South America	• Argentina • Brazil • Chile

attention to MFC activities with these countries. Instead, the methodology concentrates on the more difficult long-term prioritization decisions in conditions of deep uncertainty about the type, location, and scope of potential coalition operations. In addition, the methodology has the Army's "prepare" function in mind and does not intend to treat the international activities from the "environment-shaping" perspective that we associate primarily with the CINCs. For example, MFC activities involving Algeria may be important to EUCOM and may prove to be useful tools for engagement and environment-shaping, but the methodology presented here is not focused on the "shaping" function. Instead, the focus of this study is on MFC activities geared toward enhancing the operational effectiveness of coalitions. Finally, the composition of the priority list of countries may change as new trends emerge. The methodology presented here is flexible and can be adjusted to reflect such changes. Indeed, it is expected that the methodology will be reapplied annually. Because some propensity measures are calculated using multiyear averages (UN votes and freedom ratings), they may underestimate important events in the recent past. Army planners need to be on the alert for sudden and major shifts in country data (due, for example, to a shift from an authoritarian to a democratic regime) and weigh such considerations accordingly when preparing MFC activities. A similar recommendation applies to fairly certain future events. For example, if a given country has historically low per-capita spending on defense but embarks on a multiyear defense buildup, Army planners need to keep a close watch on that country, well before the spending figures change the average.

The specific results presented here are based on data available as of late 1999. Because of our use of multiyear averaging techniques, the results are still applicable. To be fully current for policy guidance purposes, however, later data would have to be examined.

Worldwide partners. Three major NATO states demonstrated high scores across the board in all seven categories: the United Kingdom, France, and Germany. The United Kingdom and Germany are both long-time and close allies. Although France has an ambiguous relationship with NATO's integrated command structure and is not as closely linked militarily to the United States, it has been an active and reliable participant in operations with the United States and has the capability and the demonstrated willingness that make its ground forces likely to participate in future coalition operations with the United States. Among the three countries, the United Kingdom stands out because of its considerably higher per-soldier spending levels. In terms of a worldwide future coalition partner for the United States, the United Kingdom stands out as most important in a combined measure of the quality and quantity of forces and likelihood of participation. Germany is likely to emerge as a critical world-class ally, although political constraints may hinder deployment of German troops in combat missions in the short run.

There is also a second tier of states whose high level of participation in past operations and close ties with the United States suggest a high likelihood of future participation, but whose ground force strength or per-soldier spending are at the intermediate level. The smaller potential contributors (though with high per-soldier spending) include Belgium, Canada, and the Netherlands. (Norway, whose participation has been high but whose ground troop strength is too low to expect sizable contributions, would otherwise join this group.) The second tier of countries ranked as such because of intermediate per-soldier spending (though large in size) includes Italy and Spain.

Both the first and second tier of likely coalition partners are exclusively long-standing U.S. allies. With the exception of Canada, all are also located in Europe. The only country that the methodology may have omitted unjustifiably (due to the inherent weighting bias toward Europe) from consideration for the second-tier partners, is Australia. If the bias is taken into account, then Australia joins the smaller potential contributor group in the second tier of states. Australia has intermediate-level participation and ground force strength, but it scores high in all other categories. In addition, both of the operations in which Australia participated took place outside the East Asia/Pacific region. A long-time ally, it can be expected to continue to participate in coalition operations with the United States. Consequently, we believe the case is strong for inclusion of Australia in the second tier of worldwide partners.

Regional partners. While it is clear that NATO allies will continue to be likely partners for many future operations, many other countries are likely to be involved in operations within their own regions. Thus, it is important also to look at each part of the world specifically, to identify likely partners as well as the regional "gaps" where it is difficult to predict whether any states will be likely to participate alongside the United States.

Europe. The distinguishing overall point about Europe is that most of the European states have a high or intermediate voting coincidence with the United States in the UN, almost all have a high or intermediate freedom rating, most have a high or intermediate level of U.S. ties, and an overwhelming majority have participated in some coalition operations with the United States. All of these indicators differentiate Europe from other areas of the world and, as a general area of concentration of Army MFC efforts, make Europe by far the most important continent. Thus, even though many European states rank in the category of crucial or important future coalition partners for the United States, many other European states also deserve attention from the standpoint of Army MFC priorities. Among the NATO allies (not already identified above as worldwide partners), Poland has mid-range scores for past participation and low per-soldier spending, but it also has strong ties with the United States, high UN voting and democracy ratings, substantial ground force strength, and special motivation to take part in coalition operations because of the need to prove its worth as a new ally. Among the PfP states, Romania and Bulgaria have a high level of U.S. ties, mid-range scores for past participation, low per-soldier spending, substantial troop strength in the case of Romania, and incentives to participate because of the need to enhance their position with the United States. Finally, it is important to monitor the trends that foreshadow future propensity to participate in coalition operations. Many PfP states, such as Finland, Slovakia, and Sweden, show trends toward increasing propensity to contribute over the past few years. Russia shows the opposite trends.

Middle East. Turkey is the obvious U.S. partner in the Middle East (as well as in southeastern Europe). Turkey has an intermediate ranking on voting record and democracy and a low per-soldier spending level, but it has an excellent record of participation, strong U.S. ties, and a large enough army to suggest that it can make a substantial contribution in future coalition operations. Israel's low score on previous participation and high scores on almost all other factors reflect its focus on its own military operations and its unique position in the region. While political considerations bar Israel's participation today, it could emerge quickly as a key contributor if the

Middle East peace process were to be concluded successfully. Egypt, Jordan, Morocco, and Saudi Arabia combine strong ground forces and close U.S. ties as well as some history of past participation. These states are the most important U.S. regional partners for potential coalition operations.

Africa. Most sub-Saharan countries in Africa have low or intermediate scores in all categories. Botswana is the exception, with strong U.S. ties and a good democracy rating, though a low ground strength. Given its internal transition in the 1990s, a uniquely high freedom rating in Africa since 1995, growing security ties with the United States, and regionally significant ground forces in size and sophistication, South Africa is the most likely and potentially most important sub-Saharan African contributor of forces to future regional coalition operations involving the United States. The fact that South Africa, like Nigeria, may see itself as having more of a leadership role than one that lends itself easily to partnership with the United States, however, may create difficulties and will certainly shape the form that U.S. involvement in the region will take. Many of the countries taking part in the U.S. African Crisis Response Initiative (ACRI), such as Ghana, Malawi, Mali, and Senegal, show recent trends of improvement in some indicators (freedom or UN voting ratings). This may bode well for the building of cooperation in the future.

Central and South Asia. This area is notable for the overall low scores in all categories for the vast majority of countries. For the most part, the only high scores observed reflect large ground forces. While some states have participated in operations with the United States in the past, this history seems more indicative of their high participation in UN peacekeeping as a whole, primarily for the financial gain involved. Some countries from this group may participate in coalition operations with the United States (primarily lower-end peace operations), but at this stage there is no indication of any reliable long-term partner among them.

East Asia and Pacific. Japan, South Korea, Singapore, and Thailand score high on level of ties with the United States but perform less well on most other indicators. Japan and, to a lesser extent, South Korea stand out as the most likely potential coalition partners in the future, based on UN voting coincidence and freedom ratings, though so far, for different reasons, their participation in coalition operations with the United States has been limited.[23]

North and Central America. Canada overshadows all other states in this group on the basis of its strong record of ties and cooperation with the United States and the high quality of its forces. Among the rest, El Salvador, Honduras, and Guatemala may make small contributions. Trends in Nicaragua since 1995 may mean a greater future role as a contributor.

South America. Argentina emerges as the most important and likely U.S. partner in this group on the basis of high level of U.S. ties, the most sophisticated forces on the continent, participation in past coalition operations, and the strongest record of voting coincidence with the United States in this group. Brazil and Chile are the other two important countries.

STEP 2: ESTABLISHING COMPATIBILITY DEFICIENCIES AMONG THE PARTNER COUNTRIES

The methodology to predict the propensity to participate in a coalition operation specified the most important countries for MFC-directed activities. Army planners need to add to this list the countries where a recognized threat and a U.S. commitment are in place. However, given the large differences between the forces of the countries identified as highly likely to participate in future coalition operations and U.S. forces, how can the Army structure appropriate MFC efforts toward each potential partner? The first question is to ascertain the areas where difficulties due to incompatibilities between forces may be especially problematic. We turned to DynaRank, a decision-support system (and a Microsoft Excel workbook available for Macintosh and IBM-compatible computers) designed at RAND in 1998 to assist Department of Defense decisionmaking.[24] DynaRank ranks policy options by cost-effectiveness, based on the relative

---

[23]Given its small ground force size (below the 5,000 threshold of the methodology), New Zealand was not included in the analysis.

[24]Richard J. Hillestad and Paul K. Davis, *Resource Allocation for the New Defense Strategy: The DynaRank Decision-Support System,* Santa Monica, CA: RAND, MR-996-OSD, 1998.

importance of objectives and a variety of success criteria. Although DynaRank aimed to assist the DoD's high-level resource allocation decisionmaking, its strong sensitivity to strategy, amenability to a variety of data (subjective judgments as well as quantitative analyses), and ability to link several levels of analysis makes it useful in supporting other types of defense planning.

By substantively modifying the DynaRank system, we developed the Military Compatibility Assessment Tool (MCAT). The tool aims to differentiate further between the countries determined likely to participate with the United States in coalition operations (on the basis of the analysis described above) and provides a means of assisting DUSA-IA in making strategic-level MFC plans and policy recommendations. MCAT does so by evaluating the existing level of military compatibility of select foreign ground forces relative to the U.S. Army in a variety of critical mission areas, thereby providing a rationale for MFC policy choices. MCAT ranks partners based on their ground forces' compatibility with the U.S. Army and identifies areas of greatest need and potential bottlenecks in coalition operations. It also allows for comparisons among the states while judging all against the United States as the standard point of reference. The ranking provides input to assist DUSA-IA in making decisions about which countries and which sets of capabilities (defined here as "capability assessment areas" or CAAs) need to be targeted to enhance future coalition effectiveness. Linking specific MFC assistance policies with the results from MCAT also aids the decisionmaking process for allocating resources for MFC activities. In addition, MCAT allows for variable weighting of requirements based on different operational contexts, making the tool flexible and adaptable across the mission spectrum.

We derived the MCAT criteria, both CAAs and compatibility measures (CMs), by examining multiple reports on past multinational operations and a variety of Army publications. We identified nine CAAs, which we then organized into three categories: battlefield functions (maneuver, fire support, air defense, mobility/survivability), combat service support (logistics, deployability), and C4I (C2, communications, computers, intelligence). While these criteria provide a valid and critical examination of a country's capability/compatibility, the strength of the MCAT lies in the ability to change the criteria if an examination of the underlying assumptions proves they are no longer valid or require change as the situation may dictate.[25]

In addition, three CMs stand out as most relevant: doctrine, technology/equipment, and operational readiness/training. Users of MCAT evaluate each of the CAAs along the three CMs, calculating a final overall assessment for each CAA. The CMs, drawn from a review of recent multinational force operation after-action reports and studies and current Army doctrine and regulations, represent the critical measures affecting the compatibility between two forces.[26] While compatibility problems are often attributed to equipment or technology differences between countries, MCAT takes into account the more complicated and realistic view that compatibility is based on a variety of factors.

The MCAT allows for varying the relative weights of the CAAs. If, for example, the user determines that the CAAs are unequal in their impact on compatibility, the MCAT allows for reweighting of the CAAs to represent their relative importance. The different weighting of the CAAs for MFC is most likely to vary based on mission variables such as intensity of conflict, lead time before deployment, force integration (e.g., division versus battalion), and command structure (e.g., lead nation versus parallel). Weighting different compatibility assessment areas according to their importance in different operational scenarios will allow the user to identify the relative capability of different prospective partners across different scenarios. A country with relatively capable but not easily deployable forces may score poorly if short-warning operations is the focus of analysis. This country would fare better when the emphasis is on long-lead-time missions, when its forces could be deployed using U.S. or other transportation assets. There is no single "correct" set of weighting

---

[25]For instance, Army planners seeking to identify partners' capabilities in the area of civil affairs may want to consider adding this CAA to the MCAT.

[26]Army Field Manuals FM 100-8, *The Army in Multinational Operations,* and FM 100-5, *Operations.* Also Roger H. Paulin, *Multinational Military Forces: Problems and Prospects,* London: Institute for Strategic Studies, Adelphi Paper 294, 1995; Thomas J. Marshall (ed.), with Phillip Kaiser and Jon Kessmeire, *Problems and Solutions in Future Coalition Operations,* Carlisle Barracks, PA: U.S. Army War College, Strategic Studies Institute, December 1997; Martha Maurer, *Coalition Command and Control,* Washington, D.C.: National Defense University, Institute for National Strategic Studies and the Center for Advanced Concepts and Technology, 1994.

criteria, since their determination depends on the characteristics of the partner (worldwide or regional, large or small contributor, type of likely operations in the region). In addition, multiple criteria may be necessary for some partners, especially the worldwide partners who are likely to participate in missions across the conflict spectrum. Appendix B contains a full description, including step-by-step instructions, of the MCAT. Appendix C provides a tutorial for the MCAT, using a hypothetical partner country as an example.

STEP 3: LINKING APPROPRIATE POLICIES TO IDENTIFIED COMPATIBILITY DEFICIENCIES

The next step in the process of putting together a custom-made plan of Army MFC measures consists of an accurate linking of the main deficiencies in compatibility (as established with the use of MCAT) with appropriate MFC policies available to the Army. The MCAT identifies and categorizes the shortfalls by both capability area (maneuver, combat service support, C4I) and compatibility measure (doctrine, technology, operational readiness). For instance, a country may be doctrinally compatible in its C4I doctrine, but may lack the assets to be technologically compatible.

The specific shortfalls identified in MCAT and chosen for improvement through MFC measures can then be addressed by specific MFC policies. Table 4.2 provides a sample of the types of "mitigation measures" available to the Army to address compatibility shortfalls identified by the MCAT analysis.

The identification and grouping of policies is illustrative and is based on a previous RAND study that analyzed the effects of the technology gap between the United States and its partners in coalition operations and identified certain "mitigation measures" to overcome compatibility problems.[27] The choice of specific "mitigation measures" depends greatly on the type of operation in which the other country's forces are expected to participate. In line with our earlier recommendations (in Chapter Three) on the proponent role of DUSA-IA regarding international activities designed to further MFC,

---

[27]Michele Zanini and Jennifer Morrison Taw, *The Army and Multinational Force Compatibility*, Santa Monica, CA: RAND, MR-1154-A, 2000.

Table 4.2

Framework for Deriving Mitigation Measures

Problems		Ad Hoc, High or Low Intensity, Long Lead Time
C4I	Operational	Provide C4I, liaisons; **IMET,** predeployment MTTs; **develop intel-sharing protocols**
	Organizational	Establish lead nation C2 structure, geographic separation; set up C3IC
	Technological	Loan/share/**sell** equipment; rely on lowest common denominator (LDC), COTS, SATCOM where not compromised
Logistics and Deployability	Operational	Phase deployment; provide logistics & lift; **preposition** materiel; lease lift, local transportation
	Organizational	Establish geographic separation; stovepiping
	Technological	Loan/share/sell equipment
Doctrine, Procedures and Employment	Operational	Provide liaisons; **IMET;** predeployment MTTs, **standardized** and predeployment exercises; **invite LNOs to TRADOC, War College, other Army centers;** provide missing capabilities (force protection); establish a quick reaction force
	Organizational	Establish lead nation C2 structure, geographic separation
	Technological	Loan/share/**sell** equipment; rely on COTS

NOTE: The fixes are shown in boldface type. The relative importance of fixes and workarounds changes with the amount of lead time: with short lead times, fixes become more important since some workarounds may not be feasible.

Table 4.2—extended

Ad Hoc, High or Low Intensity, Short Lead Time	Alliance, High or Low Intensity, Long or Short Lead Time
Provide C4I, liaisons; **develop intel-sharing protocols**	Provide C4I, liaison; **develop combined exercise training & intel-sharing protocols**
Establish lead nation C2 structure, geographic separation; set up C3IC	**Integrate C2 structure, forces;** partly rely on geographic separation
Loan/share/**sell** equipment; rely on LDC, COTS, SATCOM where not compromised	Loan/share/**sell** equipment; rely on LDC, **jointly develop equipment;** rely on COTS, SATCOM where not compromised
Phase deployment; provide logistics & lift; **preposition materiel;** lease local transportation	**Implement combined total asset visibility (TAV);** provide logistics & lift; **preposition materiel;** lease lift (if long lead time), local transport
Establish geographic separation; stovepiping	**Develop combined, complementary lift and logistics procedures;** or stovepipe
Loan/share/**sell** equipment	**Share, co-develop TAV; coordinate procurement to ensure compatibility**
Provide liaisons; **IMET; standardized exercises; invite LNOs to TRADOC, War College, other Army centers;** provide missing capabilities (force protection); establish a quick reaction force	**Develop combined doctrine, training, exercises, exchanges, etc.;** provide missing capabilities (force protection); **compensate in combined planning**
Establish lead nation C2 structure, geographic separation	**Integrate command structure, forces;** partly rely on geographic separation
Loan/share/**sell** equipment; rely on COTS	Loan/share/**sell** equipment; **co-develop equipment and materiel; establish compatibility protocols**

the actual selection of specific matchups between the compatibility problems identified and the "mitigation measures" available is best left to Army component commands and the Army agencies actually in charge of these measures. As long as a feedback loop to DUSA-IA exists to allow assessment of the success of these measures, a decentralized implementation of the "mitigation measures" is a better option than central control.

STEP 4: DETERMINING COST-EFFECTIVE RESOURCE ALLOCATION

The final step in the process of devising appropriate Army MFC measures would include a consideration of the budget data (program costs) to determine the cost-effectiveness of the programs. In view of the unavailability of detailed budgetary and resource data on international activities in general (as detailed in Chapter Three), step 4 is currently not possible. However, following the development of an Army MFC strategy and the identification of both MFC-enhancing policy options and program costs, the DynaRank decision support system mentioned in step 2 provides a robust and flexible tool for evaluating policy choices. In ranking policy options, DynaRank integrates consideration of program costs with both higher-level objectives and a variety of lower-level success criteria relating specifically to MFC. This decision support tool produces a hierarchy of policy choices based not only on effectiveness but also on efficiency (or "productivity").

If TEPMIS, or another tool like it, is implemented to include detailed budgetary and resource data at the program level and if the Army makes available certain additional budgetary data, the information could be used to complete a DynaRank analysis of MFC policy options. This would constitute the final step allowing detailed and verifiable planning of Army MFC efforts on the basis of effectiveness and efficiency. In line with our earlier recommendations (in Chapter Three) on the proponent role of DUSA-IA regarding international activities designed to further MFC, the actual assessment of the "productivity" of the individual MFC activities is best left to Army component commands and the Army agencies actually implementing them. However, DUSA-IA needs to be aware of the overall costs, so as to allow for a macro-level assessment of MFC policies.

Chapter Five

CONCLUSIONS

Improving the current system of planning Army MFC activities faces daunting challenges. This study has identified the following problems that hamper any cost-benefit assessment and evaluation of these efforts: (1) difficulty in obtaining input measures to assess effectiveness; (2) lack of standardized output measures of effectiveness; and (3) absence of an underlying set of goals and an Army-specific framework for MFC. Because of the basic nature of the third problem, attempting to tackle the first two without dealing with it can lead only to partial solutions. Starting from a concept of an "idealized end-state," we specified a four-step process that (1) identifies the most likely long-term U.S. coalition partners, (2) provides a means of pinpointing the compatibility shortcomings of the potential partners, (3) links specific MFC policies to the shortcomings, and (4) allows for cost-efficiency assessments at the program level and within the overall planning framework suggested here. The first three steps can be implemented immediately, whereas the fourth step is feasible in the future, if the Army modifies its organization of data that deal with its international activities. Once cost-effectiveness assessments of MFC efforts become possible, a fifth—currently hypothetical—step of integrating the Army's own force planning with that of allies and likely partners would become a realistic option. In other words, Army planners could carry out cost-benefit assessments on the basis of tradeoffs between own and ally capabilities, knowing in detail the costs involved and possibly in cooperation with select allies and partners. Such assessments potentially may lead to moving some Army capabilities to the reserve component while enhancing these capabilities among several allies.

This research effort began on the assumption that effective interactions among U.S. and partner ground forces result from successful efforts by all participants to think, plan, and allocate resources as coalition partners. As such, a high level of ground force performance results from three elements: U.S. Army capabilities developed under its Title 10 responsibilities, contributions by coalition partners, and the collective ability to interact effectively to accomplish coalition military goals. Given the near certainty that Army units will participate in a coalition framework in future operations, preparing them for cooperation with units from other armies forms an essential element of the Army's Title 10 responsibilities. The integrated framework for planning that we suggest here, especially when extended to resourcing, will allow for early planning and consideration of potential contributions by partners and thus provide the mechanism that will enhance the collective ability to function effectively in a coalition.

The methodology for predicting a given country's propensity to participate in a coalition operation that was developed as part of this study is not meant to be applied mechanistically. Certainly, the list of prioritized countries presented here does not imply abandoning international actions with the countries not identified as deserving priority. After all, the Army has limited leeway over choosing military-to-military activities with other ground forces. Not all U.S. goals served by such contacts are distinctly or uniquely related to MFC, and the Army has a duty to work to meet those objectives as well. National-level policies and guidance from the DoD and the Joint Staff constrain the choices and lead to the Army's engagement in MFC-enhancing activities as part of fulfilling such national security policies. However, a great deal of leeway exists between the usually general guidance on MFC received from JCS and the detailed plans worked out by the CINCs. The service-specific framework we suggest fits into that middle ground so as to coordinate MFC actions more effectively and take advantage of the synergies possible. In addition, the MFC framework we suggest would provide more accurate input by Army personnel into planning for coalition operations by Joint Staff and the CINCs. The introduction of TEPMIS means that information about the extent of activities of U.S. armed forces with troops from other countries will become more transparent and allow for more detailed planning. But TEPMIS provides only a tool to a goal and, without a clear goal in mind, may not reach its full potential. The prioritization of coalition partners proposed here provides that set of clear goals to guide the use of TEPMIS.

RECOMMENDATIONS

Several specific policy recommendations for the Army derive from the analysis above:

- The Army should adopt the methodology to ascertain propensity to participate in coalitions with the United States and the Military Compatibility Assessment Tool as the basic processes for prioritizing partners and developing long-range Army priorities for MFC by country and activity. HQDA can work through the Army component commanders to ensure that Army priorities are given reasonable consideration during development of the CINC theater engagement plans.
- The Program Evaluation Groups should be sensitized to the need for improved resource information to support tradeoff evaluations among activities. This need not entail the development of major accounting modifications, but rather the identification and reporting of international activity expenditures that might not be identifiable by people outside of the PEG.
- The Army should continue to participate in the development of TEPMIS, with particular attention to the process for identifying the resource implications of planned activities.
- Standardized criteria for evaluation of MFC activities should be developed matching the categorization in the MCAT. For optimum effect, further evaluation and budgeting overhauls need to take into account the MCAT.

Appendix A

COUNTRY DATA

This appendix provides a full explanation of the methodology for evaluating a country's likelihood of joining the United States in a coalition operation. It also provides the detailed data for the 108 countries that qualified for further analysis under the criteria outlined in Chapter Four.

TERMINOLOGY AND METHODOLOGY

Countries and Regions

We grouped all countries into regions, organized primarily according to geographic definitions. The list below explains the terminology in full and lists the countries considered and not considered in the survey (as well as the reasons for lack of consideration). We evaluated in detail 108 countries, grouped into seven regions:

- Europe (28)
- Central and South Asia (8)
- East Asia and Pacific (13)
- Middle East (17)
- Africa (26)
- North and Central America (7)
- South America (9)

Europe. We used the standard geographic definition: "landmass flanked by the Atlantic Ocean in the west, the Mediterranean,

Aegean, Marmara, and Black Seas and the Caucasus Mountains in the south, the Caspian Sea, Ural River, and the Ural Mountains in the east, and the Arctic Ocean in the north." We included in our grouping islands near Europe. In situations where a state spanned two continents (Russia, Turkey), we assigned it to the region where its capital was located.

We considered the following 28 states: Austria, Belarus, Belgium, Bulgaria, Croatia, Czech Republic, Denmark, Finland, France, Germany, Greece, Hungary, Ireland, Italy, Macedonia, Moldova, the Netherlands, Norway, Poland, Portugal, Romania, Russia, Slovakia, Slovenia, Spain, Sweden, Ukraine, and United Kingdom. We did not consider the following 3 states because of internal unrest: Albania, Bosnia-Herzegovina, and Cyprus. We did not consider the following state because of its position as a U.S. adversary: Yugoslavia. We did not consider the following 9 states because their ground forces were too small for our purposes: Andorra, Estonia, Iceland, Latvia, Liechtenstein, Lithuania, Luxembourg, Malta, and San Marino. We did not consider Switzerland because of its unique neutrality policy and its lack of representation in the UN.

Central and South Asia. We included in this grouping the countries of Asia ("landmass flanked by the Pacific Ocean in the east, the Indian Ocean in the south, the Red, Mediterranean, Aegean, Marmara, Black Seas, the Caucasus mountains, the Caspian Sea, Ural River, and the Ural Mountains in the west, and the Arctic Ocean in the north") but excluded the countries in the "Middle East" (see below for definition) and those in "East Asia" (see below).

We considered the following 8 states: Bangladesh, India, Kazakhstan, Kyrgyzstan, Nepal, Pakistan, Turkmenistan, and Uzbekistan. We did not consider the following 3 states because of internal unrest: Afghanistan, Sri Lanka, and Tajikistan. We did not consider the following 2 states because their ground forces were too small for our purposes: Bhutan and Maldives.

East Asia and Pacific. We included in this grouping the countries of Asia that border on the Pacific Ocean. We included Mongolia, Myanmar, and Laos because of their proximity to the Pacific. In the category of Pacific countries, we included Australia and the island states of Oceania.

We considered the following 13 states: Australia, China, Indonesia, Japan, South Korea, Laos, Malaysia, Mongolia, Myanmar, Philippines, Singapore, Thailand, and Vietnam. We did not consider the following state because of internal unrest: Cambodia. We did not consider the following state because of its position as a U.S. adversary: North Korea. We did not consider the following 10 states because their ground forces were too small for our purposes: Brunei, Fiji, Marshall Islands, Micronesia, New Zealand, Palau, Papua/New Guinea, Samoa, Solomon Islands, and Vanuatu. We did not consider Taiwan because of the dispute over its statehood and its lack of representation in the UN.

Middle East. We included in this grouping the countries of Asia west of "Central and South Asia" (see above for definition) and the countries of Africa bordering on the Mediterranean Sea.

We considered the following 17 states: Armenia, Azerbaijan, Bahrain, Egypt, Georgia, Israel, Jordan, Kuwait, Morocco, Oman, Qatar, Saudi Arabia, Syria, Tunisia, Turkey, United Arab Emirates, and Yemen. We did not consider the following 2 states because of internal unrest: Algeria and Lebanon. We did not consider the following 3 states because of their position as U.S. adversaries: Iran, Iraq, and Libya.

Africa. We included in this grouping countries located on the African continent except for those bordering on the Mediterranean Sea.

We considered the following 26 states: Botswana, Burkina Faso, Cameroon, Chad, Cote d'Ivoire, Djibouti, Eritrea, Ethiopia, Ghana, Guinea, Kenya, Madagascar, Malawi, Mali, Mauritania, Mozambique, Namibia, Niger, Nigeria, Senegal, South Africa, Tanzania, Togo, Uganda, Zambia, and Zimbabwe. We did not consider the following 10 states because of internal unrest: Angola, Burundi, Congo, Democratic Republic of Congo, Guinea-Bissau, Liberia, Rwanda, Sierra Leone, Somalia, and Sudan. We did not consider the following 12 states because their ground forces were too small for our purposes: Benin, Cape Verde, Central African Republic, Comoros, Equatorial Guinea, Gabon, Gambia, Lesotho, Mauritius, Sao Tome and Principe, Seychelles, and Swaziland.

North and Central America. We included in this grouping countries located on the American continent north of the Isthmus of Panama. We included the island states of the Caribbean Sea in this grouping.

We considered the following 7 states: Canada, Dominican Republic, El Salvador, Guatemala, Honduras, Mexico, and Nicaragua. We did not consider the following state because of internal unrest: Haiti. We did not consider the following state because of its position as a U.S. adversary: Cuba. We did not consider the following 13 states because their ground forces were too small for our purposes: Antigua and Barbuda, Bahamas, Barbados, Belize, Costa Rica, Dominica, Grenada, Jamaica, Panama, St. Kitts and Nevis, St. Lucia, St. Vincent and the Grenadines, and Trinidad and Tobago.

South America. We included in this grouping countries located on the American continent south of the Isthmus of Panama.

We considered the following 9 states: Argentina, Bolivia, Brazil, Chile, Ecuador, Paraguay, Peru, Uruguay, and Venezuela. We did not consider the following state because of internal unrest: Colombia. We did not consider the following 2 states because their ground forces were too small for our purposes: Guyana and Suriname.

Summary Tables and Color Coding

For each regional section, individual country data are preceded by a regional summary table. The summary tables indicate how well each country scores across the seven criteria identified in Chapter Four: coincidence with the United States on important and overall UN votes (see below), freedom rating (see below), participation in major military operations, ground strength, per-soldier spending, and political-military ties and alignment. The results from the summary tables are combined to draw subjective inferences of how willing and capable each nation is to engage in military operations alongside the U.S. Army. The tables summarize data by using a shading pattern (black = high, gray = medium, white = low). The ranges for each criterion are summarized in Table A.1.

Participation in military operations is limited to participation in operations Desert Storm (the Gulf War), Restore Hope (Somalia), Uphold Democracy (Haiti), Deliberate Force (Bosnia-Herzegovina),

Table A.1

Key to Summary Table Color-Coding

	Imp. Votes	Overall Votes	Freedom Rating	Number of Ops.	Ground Strength	Per-Soldier Spending	Alignment
High	0 to 33%	0 to 25%	1 to 3	4 to 7	> 60,000	> $94,000	Frequent regularized contacts, training, agreements, with U.S.
Medium	33 to 60%	25 to 45%	3.1 to 5.5	1 to 3	20,000–60,000	$19,000–$94,000	Infrequent or irregular contacts or strong ties with U.S. allies
Low	60 to 100%	45 to 100%	5.5 to 7	0	5,000–19,999	< $19,000	Few, sporadic, or no contacts

NOTE: Ground strength and defense spending figures are based on data from 1997.

Joint Endeavor (Bosnia-Herzegovina), Allied Force (Kosovo), and Joint Guardian (Kosovo). Because Joint Guardian is ongoing at the time of this writing, states participating with troops in place as of 16 September 1999 are given "credit" for participation.

Troop strength figures are rounded to the nearest 100 troops. Per-soldier spending is rounded to the nearest $1,000 and is reported in 1997 U.S. dollars.

Since UN votes and freedom rating data span 1990 to 1998, the summary tables list average values for all available years for each country. The color scheme for UN votes was obtained by assigning black to the countries whose averages ranked in the top 25 percent of the worldwide distribution for percentage of votes in common with the United States, white to those in the bottom 25 percent of the distribution, and gray to scores between the 25th and 75th percentiles. Separating countries in this fashion is a more coherent approach than assigning arbitrary cutoffs based on actual percentage values, given the uneven distribution of voting averages.[1] Freedom rating

[1]For instance, few countries vote with more than 60 percent coincidence with the United States on overall votes. Had the boundaries been set at 0 to 33 percent for white, 33.1 to 66 percent for gray, and 66.1 to 100 percent for black, a vast majority of countries would have ended up as "gray," with only a very small number of nations as "black" or "white." In turn, this would have made any meaningful comparison between countries difficult.

color schemes are based on the cutoffs provided by the Freedom House.

The ranges chosen for number of operations are uneven: white for zero, gray for one to three operations, and black for four to seven. This reflects the subjective judgment that a country's participation in one major operation in recent years is significantly better than no participation at all. It also reflects the assumption that only those countries that participated in more than half of the operations identified should be ranked as "black." Ground-strength low, medium, and high cutoffs are the assessed ability to contribute to multinational operations at a strength of approximately one brigade, one division, or more, respectively. Defense-spending low, medium, and high cutoffs are the percentage of U.S. per-soldier spending: 20 percent or less, 20–50 percent, and over 50 percent, respectively. Again, these ranges are arbitrarily set and should be viewed as approximations. They are nonetheless reasonable standards for a first-cut comparison across countries.

UN General Assembly Votes

Data for voting coincidence in the United Nations are from the yearly reports prepared by the State Department. These reports assess the voting practices of the governments of UN member states in the General Assembly, and they evaluate the actions and responsiveness of those governments to U.S. policy on issues of special importance to the United States.

Two data points are presented for each country from 1990 to 1998: voting coincidence with U.S. votes in overall General Assembly actions (excluding consensus resolutions, which are passed without a vote), and coincidence on important votes.[2] The State Department identifies important votes as those that directly affect important U.S. interests and on which the United States lobbies extensively.

Voting coincidence is the percentage of the total votes that a particular country voted with the United States. It is calculated by dividing the number of identical votes by the total number of overall and

[2]The large majority of resolutions passed in the General Assembly are based on consensus and do not require a vote.

important votes. It is a conservative measure of voting affinity, since it excludes abstentions and does not adjust for differences in participation (some states may not be present when votes are taken).

UN Votes: Individual Country Remarks

- Armenia admitted as member in the 1992 session.
- Dominican Republic: 0 percent for all votes in 1994 and 1995 (absent for all votes).
- Eritrea admitted as member in the 1994 session.
- Zimbabwe and Uganda in 1998: 0 percent on important votes.
- 1990 votes for Belarus and Ukraine reflect their status at that time as Soviet Socialist Republics (as a result, there is no freedom rating for that year).
- 1990 and 1991 votes for Russia are of the USSR.
- Slovenia and Croatia admitted as members in the 1992 session.
- Czech Republic and Slovakia admitted in 1993.
- The former Soviet republics of Azerbaijan, Georgia, Kazakhstan, Turkmenistan, Kyrgyzstan, and Uzbekistan did not begin to vote until 1992. Uzbekistan's and Georgia's 0 percent coincidence in 1992 has to do with these countries' absence from the General Assembly. Turkmenistan and Kyrgyzstan were also absent for several votes in the same period.
- South Korea was not a member of the UN until 1990.

Freedom Ratings

The Freedom House conducts yearly assessments of freedom status around the world. For each country, scores on a scale of 1 (free) to 7 (not free) are assigned for political rights and civil liberties. A country's overall freedom status is obtained by computing the average of the two freedom indicators. Countries whose combined averages for political rights and for civil liberties fall between 1.0 and 2.5 are designated "free"; between 3.0 and 5.5 "partly free"; and between 5.5 and 7.0 "not free."[3]

Several countries became independent, split into two or more countries, or merged with a neighbor between 1990 and 1998. Scores for these countries are given only for the period of their existence as independent states. For some countries, such as Slovenia, freedom ratings were assigned in the year prior to full UN membership, and are recorded accordingly.

---

[3]Methodological changes have been effected periodically by the Freedom House. For discussions of these, consult the methodological essays for various survey editions, particularly those for 1989–90, 1993–94, and 1995–96.

Europe (28 countries)

	Important Votes	Overall Votes	Freedom Rating	Number of Ops	Ground Strength	Per-Soldier Spending	Ties & Alignment
Austria	0.67	0.45	1.00	2	45,500	$39,000–$49,000	Medium
Belarus	0.44	0.30	4.88	0	50,500	$4,000–5,000	Low
Belgium	0.69	0.53	1.17	5	28,500	$85,000–$104,000	High
Bulgaria	0.64	0.46	2.44	1	50,400	$2,000–$3,000	High
Croatia	0.58	0.44	3.94	0	50,000	$20,000	Low
Cz. Republic	0.69	0.56	1.50	2	27,000	$15,000–$16,000	High
Denmark	0.70	0.49	1.00	3	19,000	$86,000–$102,000	High
Finland	0.68	0.48	1.00	2	27,000	$63,000–$73,000	Medium
France	0.72	0.57	1.50	6	219,900	$109,000–$130,000	High
Germany	0.70	0.53	1.50	6	240,000	$96,000–$119,000	High
Greece	0.57	0.40	1.83	3	116,000	$34,000–$37,000	High
Hungary	0.69	0.49	1.67	2	31,600	$11,000–$14,000	High
Ireland	0.68	0.45	1.11	2	10,500	$60,000–$63,000	Low
Italy	0.70	0.53	1.44	6	188,300	$67,000–$70,000	High
Macedonia	0.57	0.49	3.36	0	15,400	$9,000	Medium
Moldova	0.57	0.46	4.06	0	9,300	$2,000–$5,000	Medium
Netherlands	0.71	0.54	1.00	5		$120,000–$148,000	High
Norway	0.69	0.48	1.00	4	15,800	$99,000–$113,000	High
Poland	0.69	0.47	1.78	3	168,700	$13,000	High
Portugal	0.69	0.50	1.06	3	32,100	$43,000–$50,000	High
Romania	0.66	0.47	3.56	2	129,400	$3,000	High
Russia	0.54	0.38	3.50	2	420,000	$20,000–$52,000	Medium
Slovakia	0.66	0.54	2.75	1	23,800	$10,000–$11,000	Medium
Slovenia	0.66	0.49	1.69	0	9,600	$32,000	High
Spain	0.65	0.47	1.33	4	128,500	$39,000–$45,000	High
Sweden	0.90	0.46	1.00	3	35,100	$103,000–$116,000	Medium
Ukraine	0.50	0.35	3.44	2	161,500	$3,000–$4,000	Medium
UK	0.72	0.64	1.50	6	112,200	$162,000–$167,000	High

NOTE: The first three columns report average values, 1990–1998.

KEY	Imp. Votes[a]	Overall Votes[a]	Freedom Rating	Number of Ops.	Ground Strength[b]	Per-Soldier Spending[c]	Alignment
High	0 to 33%	0 to 25%	1 to 3	4 to 7	> 60,000	>$94,000	Frequent regularized contacts, training, agreements, with U.S.
Medium	33 to 60%	25 to 45%	3.1 to 5.5	1 to 3	20,000–60,000	$19,000–$94,000	Infrequent or irregular contacts **or** strong ties with U.S. allies
Low	60 to 100%	45 to 100%	5.5 to 7	0	5,000–19,999	<$19,000	Few, sporadic, or no contacts

[a]For UN votes, the high, medium, and low range cutoffs represent the 0–25th, 25–75th, and 75–100th percentiles, respectively.

[b]Ground-strength low, medium, and high cutoffs are the assessed ability to contribute to multinational operations at a strength of approximately one brigade, one division, or more, respectively. Based on 1997 data.

[c]Defense-spending high, medium, and low cutoffs are the percentage of U.S. per-soldier spending at 20% or less, 20–50%, and over 50%, respectively. Based on 1997 data.

Austria

UN voting coincidence and freedom rating

RANDMR1291/austria

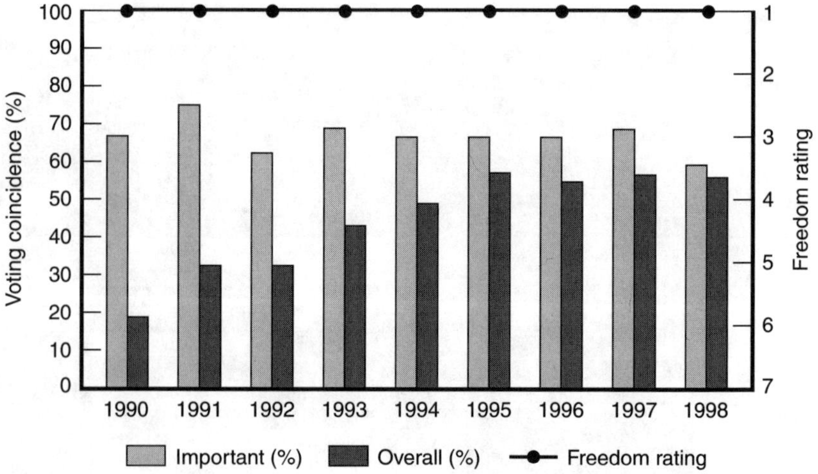

Participation in recent major operations

Desert Storm	Restore Hope	Uphold Democracy	Deliberate Force	Joint Endeavor	Allied Force	Joint Guardian
				✓		✓

Force size and defense spending

Ground force size	Total force size	Defense spending per soldier
45,500	45,500	approximately $39,000–$49,000

Alignment

Military:

- WEU observer.
- PfP member, but participation is limited to peacekeeping and humanitarian exercises.
- Has also participated in noncombat NATO exercises in the past.

Nonmilitary:

- EU member.

Defense and military agreements with the United States

- 1957, agreement on procurement by Austria of U.S. military equipment, services, etc.

Defense ties with the United States

Ties through PfP and associated peacekeeping/humanitarian cooperation.

Belarus

UN voting coincidence and freedom rating

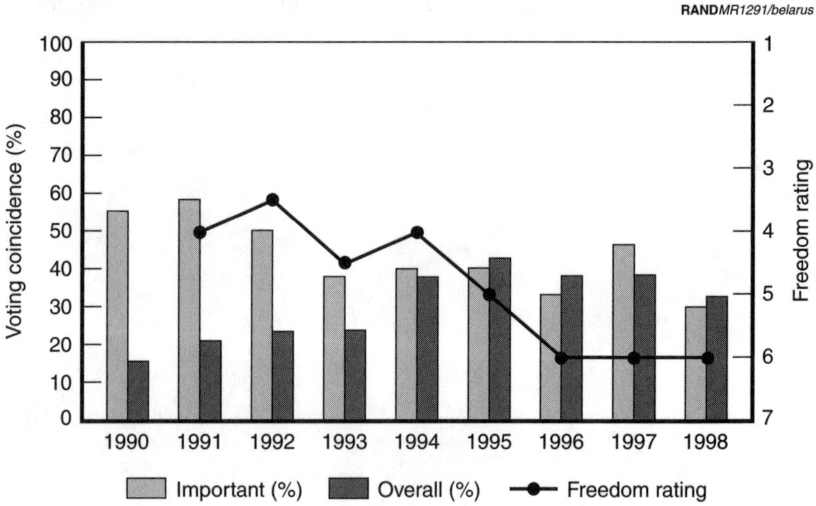

Participation in recent major operations

Desert Storm	Restore Hope	Uphold Democracy	Deliberate Force	Joint Endeavor	Allied Force	Joint Guardian

Force size and defense spending

Ground force size	Total force size	Defense spending per soldier
50,500	81,800	approximately $4,000–5,000

Alignment

Military:

- PfP (participation level low).
- CIS (with military cooperation provisions).
- Close ties with Russia.
- May be selling weapons to Iran and Syria.
- Chinese cadets training at Belarus military academy.

Defense and military agreements with the United States

- Range of cooperative threat reduction (CTR—nuclear, chemical, and biological weapons dismantlement assistance) agreements beginning in 1992.

Defense ties with the United States

Through PfP, CTR. Relations have deteriorated since President Lukashenka took power in 1994.

Belgium

UN voting coincidence and freedom rating

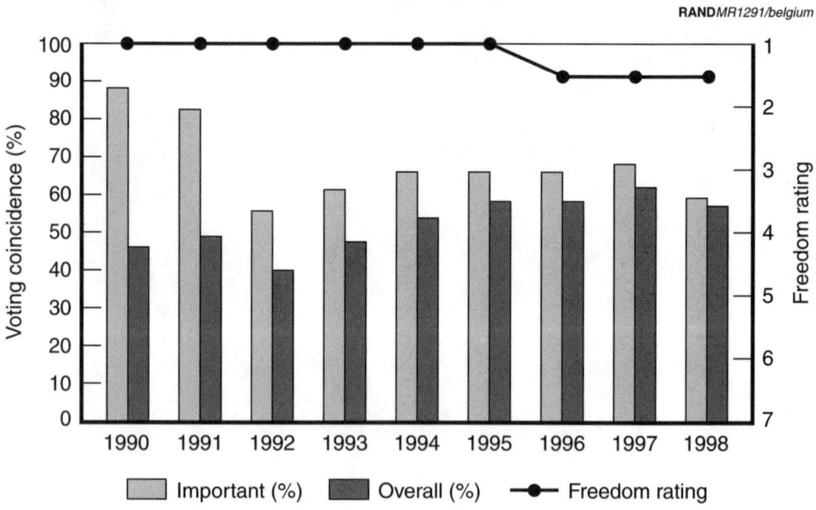

Participation in recent major operations

Desert Storm	Restore Hope	Uphold Democracy	Deliberate Force	Joint Endeavor	Allied Force	Joint Guardian
✓	✓			✓	✓	✓

Force size and defense spending

Ground force size	Total force size	Defense spending per soldier
28,500	44,500	approximately $85,000–$104,000

Alignment

Military:

- NATO.
- WEU (including European Corps).
- One weak mechanized brigade permanently deployed to Germany.
- U.S. forces (but not major units) in country.
- Benelux Deployable Air Task Force.
- British logistics base at Antwerp.

- Multinational Division (Central) with Germany, U.K., Netherlands.

Nonmilitary:

- EU.

Defense and military agreements with the United States

- 1950, mutual defense assistance.
- 1953–1984, range of agreements governing weapons production; training and personnel exchanges; other cooperation, etc.

Defense ties with the United States

Strong bilateral relationship as well as close ties through NATO.

Bulgaria

UN voting coincidence and freedom rating

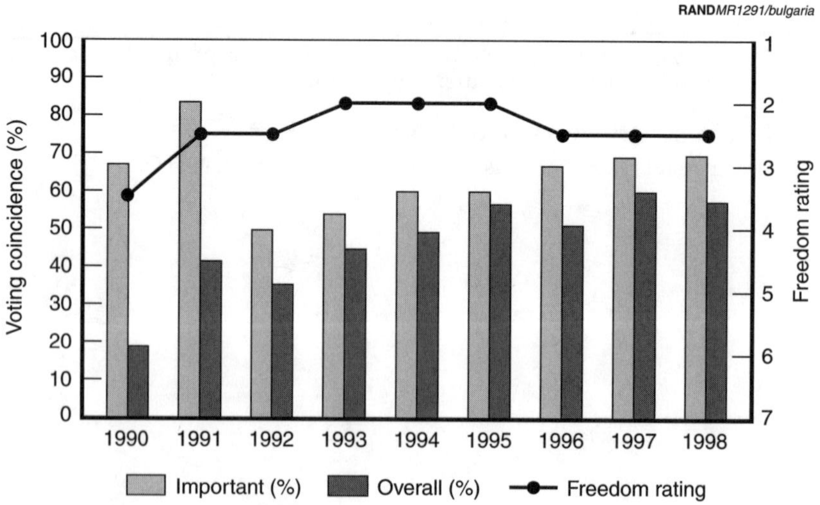

Participation in recent major operations

Desert Storm	Restore Hope	Uphold Democracy	Deliberate Force	Joint Endeavor	Allied Force	Joint Guardian
				✓		

Force size and defense spending

Ground force size	Total force size	Defense spending per soldier
50,400	101,500	$2,000–$3,000

Alignment

Military:

- PfP.
- WEU associate partner.
- Some officers trained in U.K., Germany, France, Italy, U.S.
- Military cooperation agreements with Switzerland, Russia, Albania, Austria, France, Germany, Romania.
- Informal mutual defense arrangements with Greece, Turkey.

- Southeast European Brigade with Albania, Italy, Romania, Greece, Macedonia, Turkey.

Nonmilitary:

- EU applicant with Europe Association Agreement.

Defense and military agreements with the United States

- 1992, IMET.
- 1994, military assistance (Foreign Assistance Act).
- 1995, protection of classified information.
- 1998, Bilateral Assistance Treaty.

Defense ties with the United States

Has hosted PfP exercises and been an active partner.

Croatia

UN voting coincidence and freedom rating

RANDMR1291/croatia

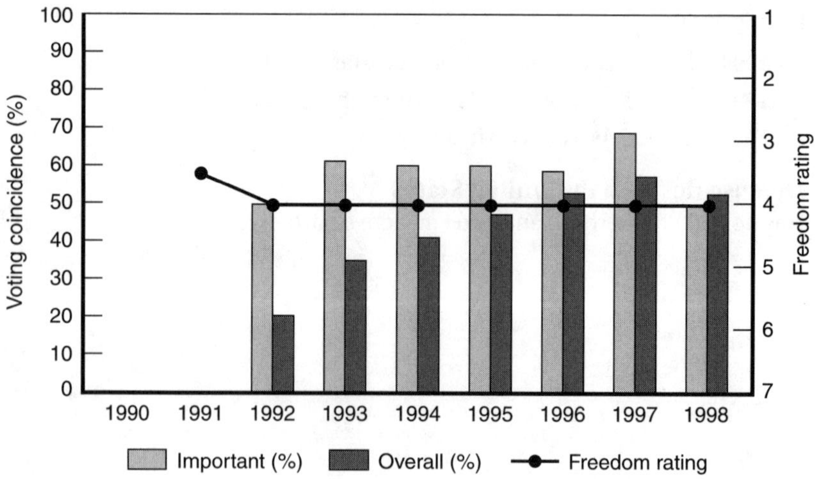

Participation in recent major operations

Desert Storm	Restore Hope	Uphold Democracy	Deliberate Force	Joint Endeavor	Allied Force	Joint Guardian

Force size and defense spending

Ground force size	Total force size	Defense spending per soldier
50,000	58,000	approximately $20,000

Alignment

Military:

- Has asked to join PfP.
- NATO South-East Europe consultative forum.

Nonmilitary:

- Hopes to begin membership talks with the EU.

Defense ties with the United States

Diplomatic ties focused on issues related to the former Yugoslavia.

Czech Republic

UN voting coincidence and freedom rating

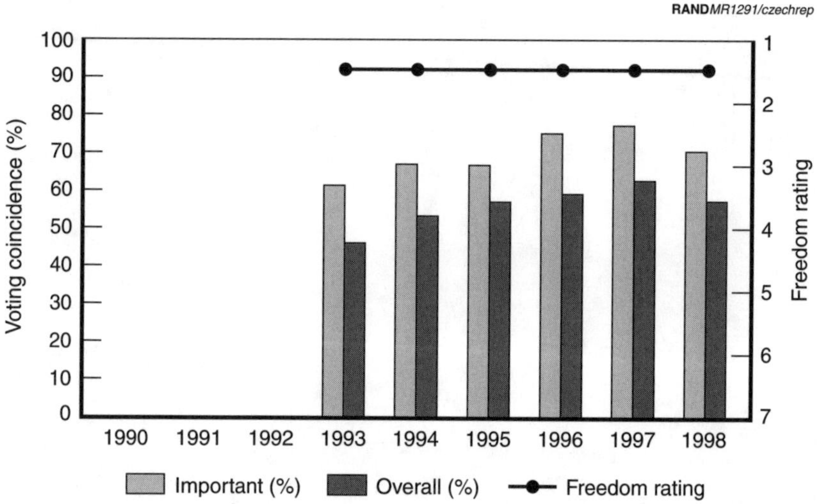

Participation in recent major operations

Desert Storm	Restore Hope	Uphold Democracy	Deliberate Force	Joint Endeavor	Allied Force	Joint Guardian
*				✓		✓

*Czechoslovakia participated in Desert Storm with a small contingent, but its deployment is not counted here.

Force size and defense spending

Ground force size	Total force size	Defense spending per soldier
27,000	61,700	approximately $15,000–$16,000

Alignment

Military:

- NATO (new).
- WEU associate member.

Nonmilitary:

- EU applicant with Europe Association Agreement.

Defense and military agreements with the United States

- 1995, protection of classified information.
- 1996, information exchange.
- 1996, cross-servicing and acquisition.

Defense ties with the United States

NATO ties.

Denmark

UN voting coincidence and freedom rating

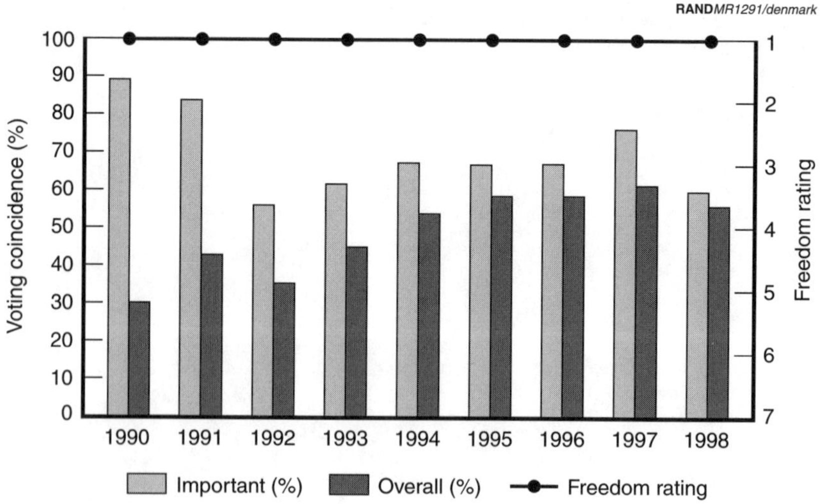

Participation in recent major operations

Desert Storm	Restore Hope	Uphold Democracy	Deliberate Force	Joint Endeavor	Allied Force	Joint Guardian
				✓	✓	✓

Force size and defense spending

Ground force size	Total force size	Defense spending per soldier
19,000	32,900	approximately $86,000–102,000

Alignment

Military:

- NATO.
- WEU observer.
- Combined Corps with Poland, Germany.
- Helped equip Baltic Battalion, which served with Danish units in SFOR.
- Formed Reaction Brigade to improve compatibility with NATO concept.
- NORDBAT (Nordic Battalion with Sweden and Norway).

- Funding peacekeeping center in Zimbabwe.

Nonmilitary:

- EU.

Defense and military agreements with the United States

- 1950, mutual defense assistance.
- 1951, Greenland defense.
- 1956, SOFA.
- Range of similar and related agreements governing weapons production; training and personnel exchanges; other cooperation, etc. from 1950–1991.

Defense ties with the United States

Close ties and cooperation through NATO, including operations and exercises.

Finland

UN voting coincidence and freedom rating

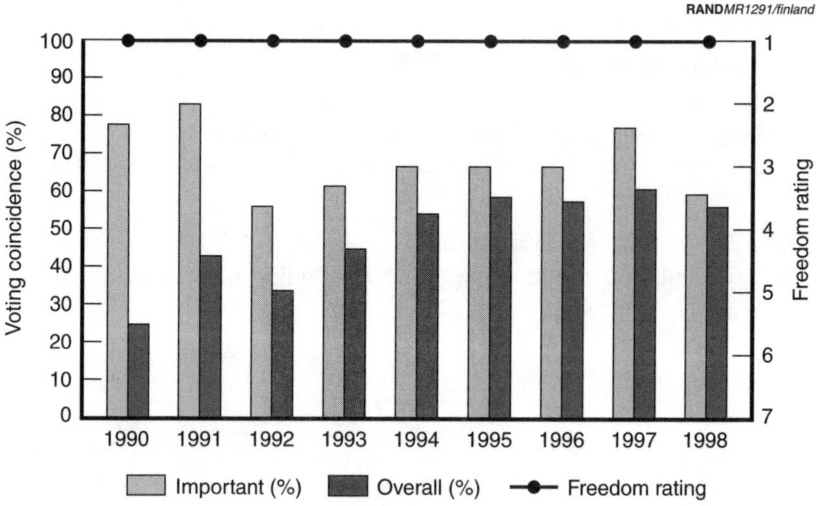

Participation in recent major operations

Desert Storm	Restore Hope	Uphold Democracy	Deliberate Force	Joint Endeavor	Allied Force	Joint Guardian
				✓		✓

Force size and defense spending

Ground force size	Total force size	Defense spending per soldier
27,000	31,000	approximately $63,000–$73,000

Alignment

Military:

- PfP.
- WEU observer.
- MOU with U.K. on defense cooperation and bilateral relations.

Nonmilitary:

- EU.

Defense and military agreements with the United States

- 1991, security of military information.
- 1995, research exchange.

Defense ties with the United States

Through PfP, SFOR.

France

UN voting coincidence and freedom rating

RANDMR1291/france

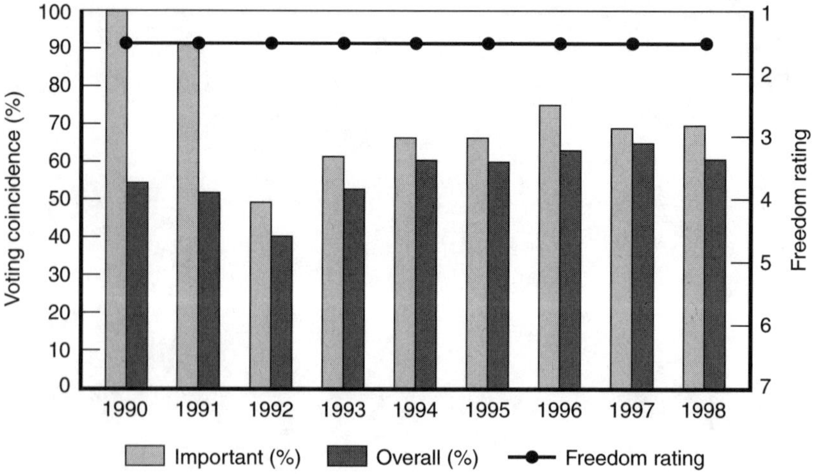

Participation in recent major operations

Desert Storm	Restore Hope	Uphold Democracy	Deliberate Force	Joint Endeavor	Allied Force	Joint Guardian
✓	✓		✓	✓	✓	✓

Force size and defense spending

Ground force size	Total force size	Defense spending per soldier
219,900	380,800	approximately $109,000–$130,000

Alignment

Military:

- Manila Pact (SEATO mutual defense agreement, still in force).
- NATO (not in integrated military command).
- WEU (including European Corps).
- High involvement in other regional military formations.
- Joint Franco-German Weapons Agency.
- 1997, exercise with Burkina Faso, Benin, Togo.

- 1997, exercise with Mauritius, Madagascar, Mayotte, Djibouti, South Africa.
- Reportedly, significant influence over Central African Republic.
- Bilateral Security Pact with Cameroon.
- Works with U.S., U.K. to build African Peacekeeping Force.

Nonmilitary:

- EU.

Defense and military agreements with the United States

- 1950, mutual defense assistance agreement.
- Range of agreements from 1950–1995 covering bases, facilities, security of information, procurement, cooperative research, scientist exchange, etc.

Defense ties with the United States

NATO; participates in exercises and general cooperation program, including considerable high-level interaction.

Germany

UN voting coincidence and freedom rating

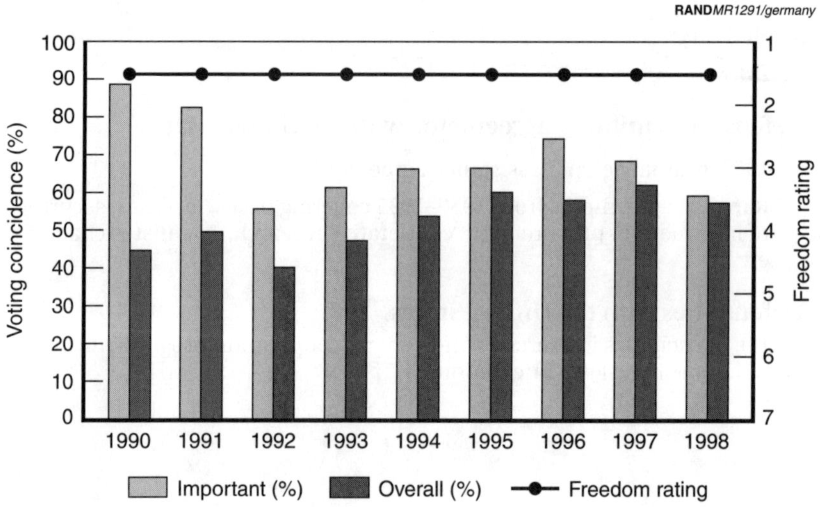

Participation in recent major operations

Desert Storm	Restore Hope	Uphold Democracy	Deliberate Force	Joint Endeavor	Allied Force	Joint Guardian
✓	✓		✓	✓	✓	✓

Force size and defense spending

Ground force size	Total force size	Defense spending per soldier
240,000	347,100	approximately $96,000–$119,000

Alignment

Military:

- NATO.
- WEU (including European Corps).
- Combined Corps with Poland, Denmark.
- Franco-German Brigade in European Corps.
- High involvement in other regional military formations.
- Some training in Chile.

Nonmilitary:

- EU.

Defense and military agreements with the United States

- 1955, mutual defense assistance.
- 1953–present, range of agreements covering bases, equipment; weapon procurement; training; etc.

Defense ties with the United States

Strong U.S. forward presence in Germany. High level of cooperation bilaterally and within NATO.

Greece

UN voting coincidence and freedom rating

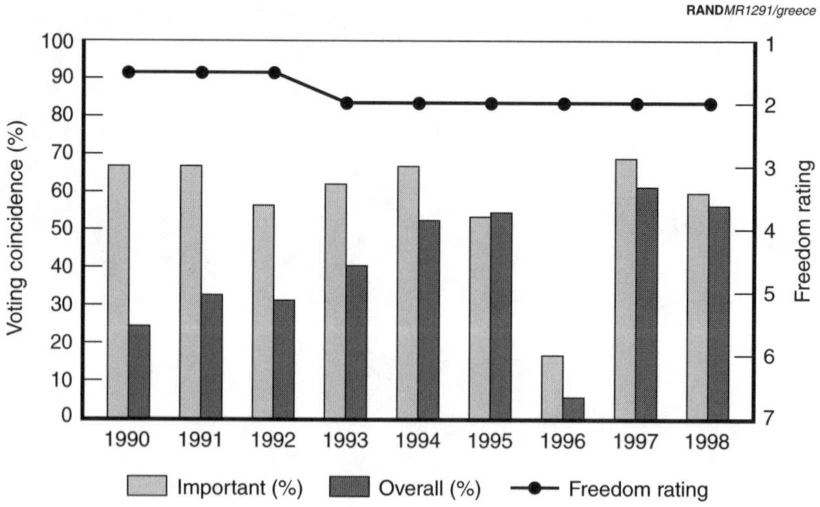

Participation in recent major operations

Desert Storm	Restore Hope	Uphold Democracy	Deliberate Force	Joint Endeavor	Allied Force	Joint Guardian
	✓			✓		✓

Force size and defense spending

Ground force size	Total force size	Defense spending per soldier
116,000	162,300	approximately $34,000–$37,000

Alignment

Military:

- NATO, but very little training with other NATO states.
- WEU.
- Southeast European Brigade with Albania, Bulgaria, Romania, Italy, Macedonia, Turkey.
- Defense Cooperation Agreement with Israel.
- Defense Cooperation Agreement with Egypt.

- Treaty of Peace and Friendship with Albania.

Nonmilitary:

- EU.

Defense and military agreements with the United States

- 1940–present, range of agreements and arrangements governing use of land and facilities, equipment and materiel, SOFA, military assistance, cooperation, etc.
- 1990, mutual defense cooperation.

Defense ties with the United States

U.S. presence; NATO ties.

Hungary

UN voting coincidence and freedom rating

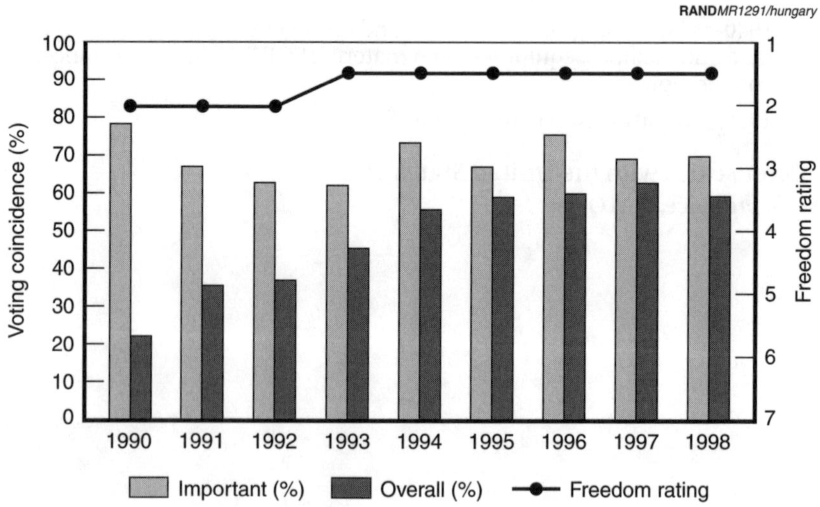

Participation in recent major operations

Desert Storm	Restore Hope	Uphold Democracy	Deliberate Force	Joint Endeavor	Allied Force	Joint Guardian
				✓		✓

Force size and defense spending

Ground force size	Total force size	Defense spending per soldier
31,600	49,100	approximately $11,000–$14,000

Alignment

Military:

- NATO (new).
- WEU associate member.
- Slovenian-Hungarian-Italian Brigade.

Nonmilitary:

- EU applicant with Europe Association Agreement.

Defense and military agreements with the United States

- 1995, scientific data exchange.
- 1996, protection of classified information.
- 1996, logistic support.

Defense ties with the United States

Exercises and activities in NATO framework.

Ireland

UN voting coincidence and freedom rating

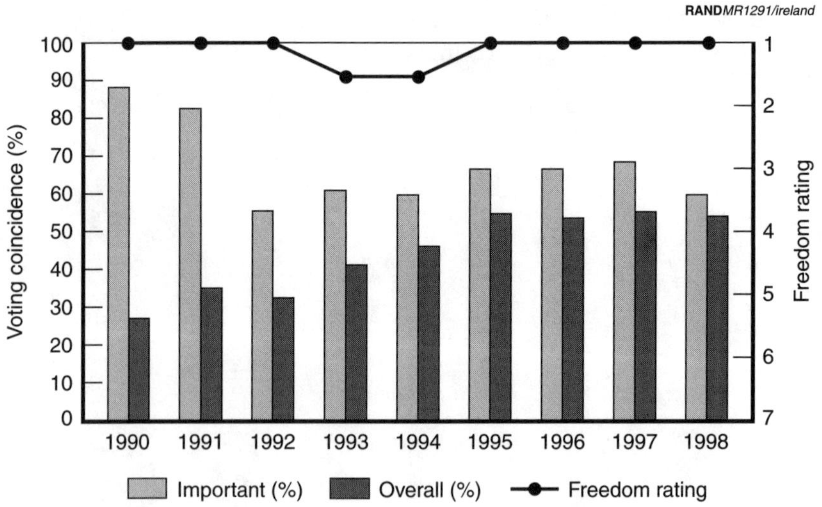

Participation in recent major operations

Desert Storm	Restore Hope	Uphold Democracy	Deliberate Force	Joint Endeavor	Allied Force	Joint Guardian
				✓ (SFOR)		✓

Force size and defense spending

Ground force size	Total force size	Defense spending per soldier
10,500	12,700	approximately $60,000–$63,000

Alignment

Military:

- WEU observer.

Nonmilitary:

- EU.

Italy

UN voting coincidence and freedom rating

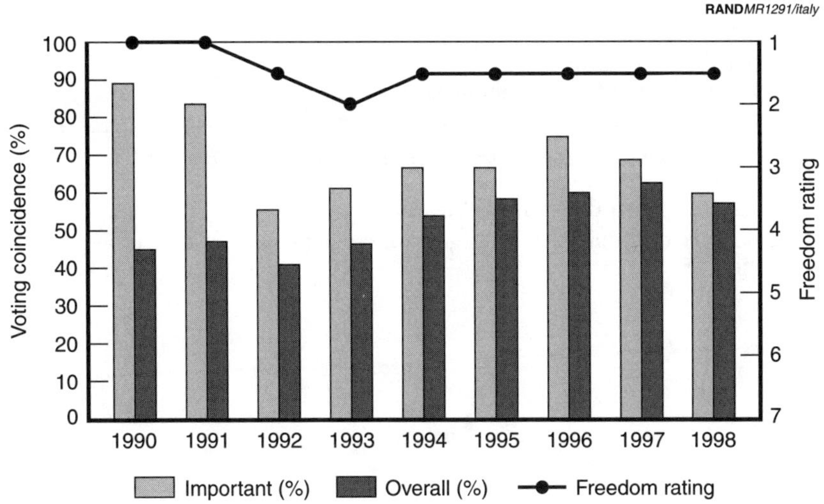

Participation in recent major operations

Desert Storm	Restore Hope	Uphold Democracy	Deliberate Force	Joint Endeavor	Allied Force	Joint Guardian
✓	✓		✓	✓	✓	✓

Force size and defense spending

Ground force size	Total force size	Defense spending per soldier
188,300	325,200	approximately $67,000–$70,000

Alignment

Military:

- NATO.
- WEU.
- Joint Amphibious Brigade with Spain.
- Eurofor with France, Spain, Portugal.
- Supports Slovenia for NATO membership.
- Slovenian-Hungarian-Italian Brigade.

- Southeast European Brigade with Albania, Bulgaria, Romania, Greece, Macedonia, Turkey.
- European Air Corps with U.K., France, Germany.

Nonmilitary:

- EU.

Defense and military agreements with the United States

- 1950, mutual defense assistance.
- Range of facilities, weapon production, classified information, logistical support, officer exchange, etc. agreements through present.

Defense ties with the United States

U.S./NATO troops based in Italy, general range of alliance and some bilateral activities.

Macedonia

UN voting coincidence and freedom rating

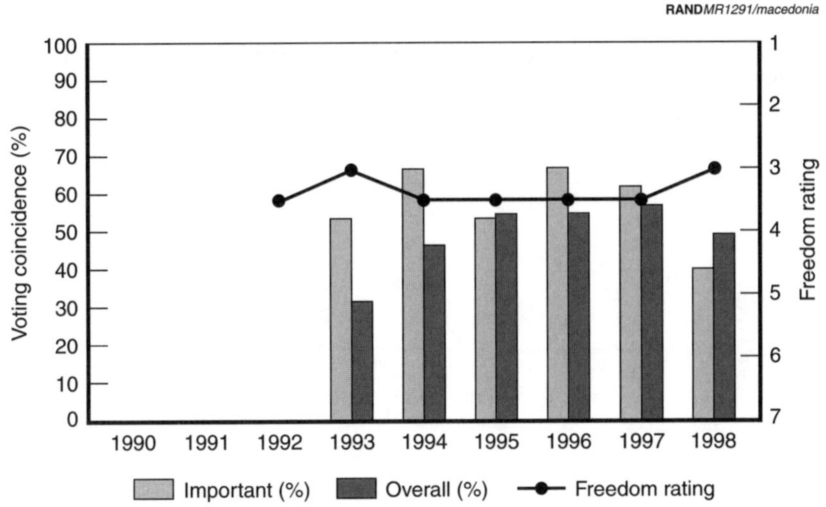

Participation in recent major operations

Desert Storm	Restore Hope	Uphold Democracy	Deliberate Force	Joint Endeavor	Allied Force	Joint Guardian

Force size and defense spending

Ground force size	Total force size	Defense spending per soldier
15,400	15,400	approximately $9,000

Alignment

Military:

- PfP.
- Improved ties with Bulgaria.
- Five-year agreement on strengthening military ties signed with Albania in 1998.

- Southeast European Brigade with Albania, Bulgaria, Romania, Greece, Italy, Turkey.

Defense and military agreements with the United States

- NATO SOFA applies.

Defense ties with the United States

PfP; U.S./NATO forces on ground in conjunction with former Yugoslavia operations.

Moldova

UN voting coincidence and freedom rating

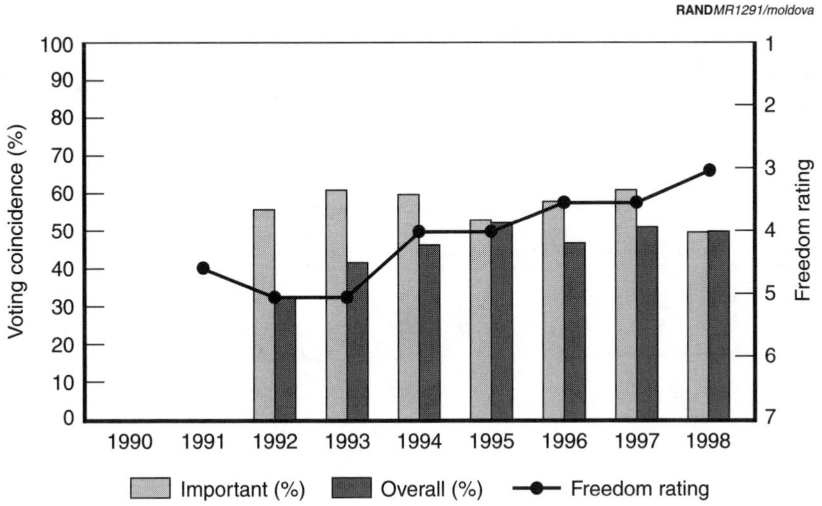

Participation in recent major operations

Desert Storm	Restore Hope	Uphold Democracy	Deliberate Force	Joint Endeavor	Allied Force	Joint Guardian

Force size and defense spending

Ground force size	Total force size	Defense spending per soldier
9,300	11,000	approximately $2,000–$5,000

Alignment

Military:

- PfP, including hosting exercises.
- GUUAM.
- Defense Cooperation Agreement with Romania.

Defense and military agreements with the United States

- 1996, NATO/PfP SOFA declared to apply.

- 1997, military assistance (Foreign Assistance Act) and furnishing of defense articles, related training, and other defense services.
- 1997, nonproliferation and promotion of defense and military relations (cooperative threat reduction).

Defense ties with the United States

PfP, cooperative threat reduction.

Netherlands

UN voting coincidence and freedom rating

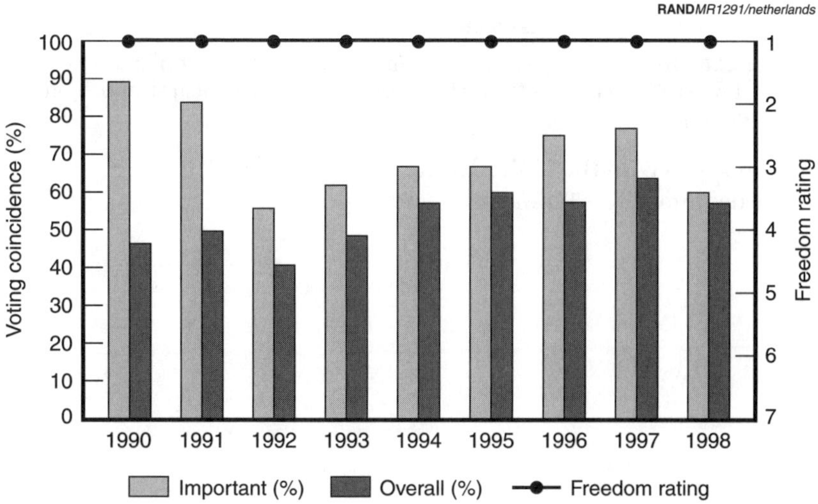

Participation in recent major operations

Desert Storm	Restore Hope	Uphold Democracy	Deliberate Force	Joint Endeavor	Allied Force	Joint Guardian
		✓	✓	✓	✓	✓

Force size and defense spending

Ground force size	Total force size	Defense spending per soldier
27,000	57,200	approximately $120,000–$148,000

Alignment

Military:

- NATO.
- WEU.
- Multinational Division (Central) with U.K., Germany, Belgium.
- Netherlands-German Corps.
- Benelux Deployable Air Task Force.

- U.K.-Netherlands Amphibious Force.

Nonmilitary:

- EU.

Defense and military agreements with the United States

- 1950, mutual defense assistance.
- Agreements through present covering Dutch participation in Korea, SOFA, weapons production, M109 cooperation, logistical support, officer exchange, etc.

Defense ties with the United States

Close ties through NATO and bilaterally.

Norway

UN voting coincidence and freedom rating

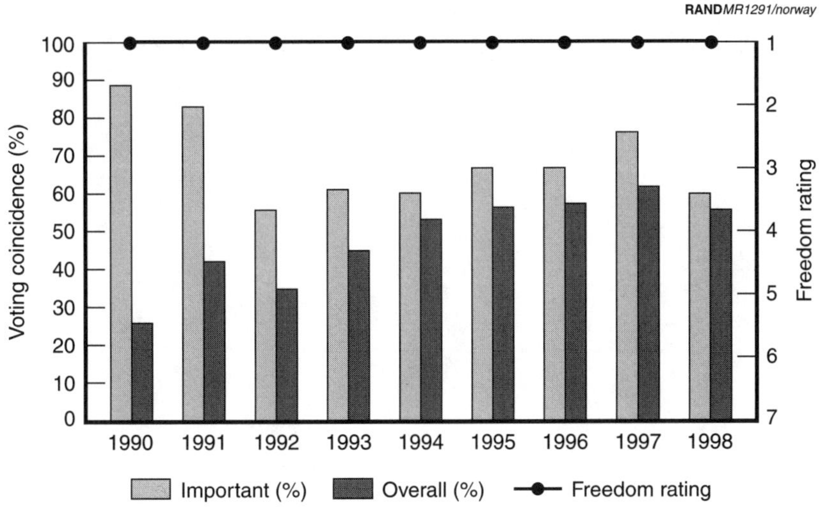

Participation in recent major operations

Desert Storm	Restore Hope	Uphold Democracy	Deliberate Force	Joint Endeavor	Allied Force	Joint Guardian
	✓			✓	✓	✓

Force size and defense spending

Ground force size	Total force size	Defense spending per soldier
15,800	33,600	approximately $99,000–$113,000

Alignment

Military:

- NATO.
- WEU associate member.
- Small U.K. presence.
- NORDBAT (Nordic Battalion with Denmark and Sweden).

Defense and military agreements with the United States

- 1950, mutual defense assistance.
- Agreements through present cover classified information, co-production, R&D, logistical support, officer exchange, etc.

Defense ties with the United States

Through NATO and bilateral. U.S. conducts exercises in Norway.

Poland

UN voting coincidence and freedom rating

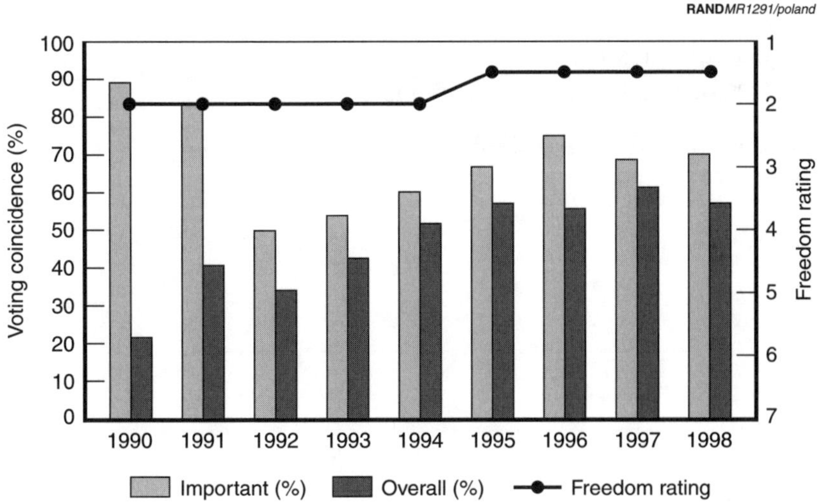

Participation in recent major operations

Desert Storm	Restore Hope	Uphold Democracy	Deliberate Force	Joint Endeavor	Allied Force	Joint Guardian
✓				✓		✓

Force size and defense spending

Ground force size	Total force size	Defense spending per soldier
168,700	241,800	approximately $13,000

Alignment

Military:

- NATO (new).
- WEU associate member.
- Combined Corps with Denmark, Germany.
- Joint Peacekeeping battalion with Ukraine.
- Joint Peacekeeping battalion with Lithuania.

Nonmilitary:

- EU applicant with Europe Association Agreement.

Defense and military agreements with the United States

- 1991, IMET.
- 1995, protection of classified information.
- 1995, exchange of R&D.
- 1996, acquisition and cross-servicing.

Defense ties with the United States

NATO ties.

Portugal

UN voting coincidence and freedom rating

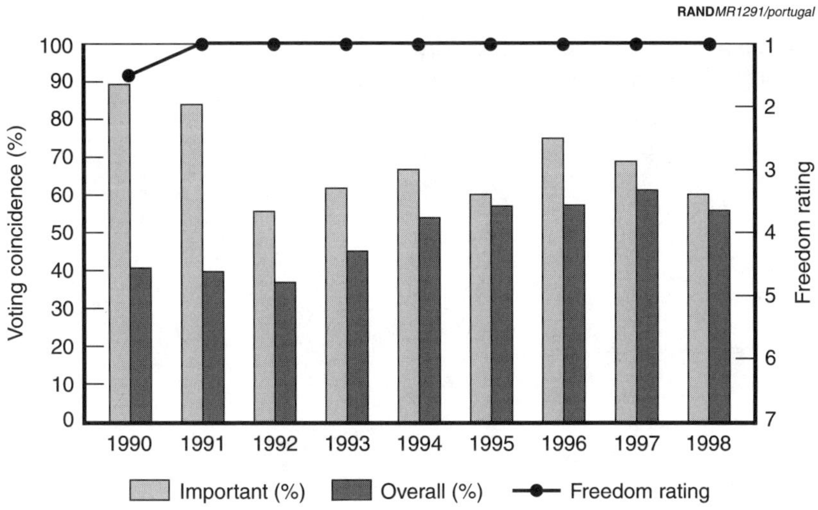

Participation in recent major operations

Desert Storm	Restore Hope	Uphold Democracy	Deliberate Force	Joint Endeavor	Allied Force	Joint Guardian
				✓	✓	✓

Force size and defense spending

Ground force size	Total force size	Defense spending per soldier
32,100	59,300	approximately $43,000–$50,000

Alignment

Military:

- NATO.
- WEU.
- Eurofor with France, Italy, Spain.

Nonmilitary:

- EU.

Defense and military agreements with the United States

- 1951, mutual defense assistance.
- 1951, Azores.
- Range of agreements and arrangements through present covering weapons production, military assistance, cooperation, etc.

Defense ties with the United States

Strong NATO and bilateral defense cooperation program.

Romania

UN voting coincidence and freedom rating

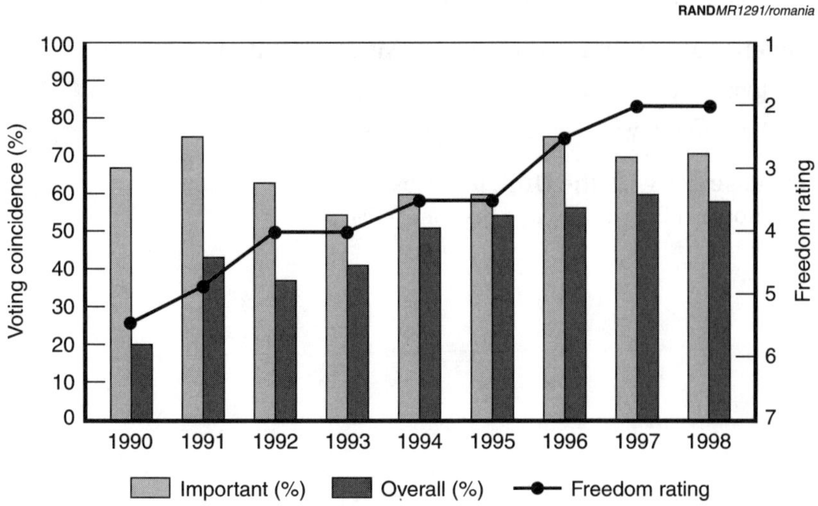

Participation in recent major operations

Desert Storm	Restore Hope	Uphold Democracy	Deliberate Force	Joint Endeavor	Allied Force	Joint Guardian
✓				✓		

Force size and defense spending

Ground force size	Total force size	Defense spending per soldier
129,400	227,000	approximately $3,000

Alignment

Military:

- PfP.
- WEU associate partner.
- Southeast European Brigade with Albania, Bulgaria, Italy, Greece, Macedonia, Turkey.
- Has offered NATO use of territory and ranges for maneuver and live-fire training.

- France supports NATO membership.
- June 1997, defense cooperation agreement with Moldova (training, joint exercises, formation of joint peacekeeping battalion).

Nonmilitary:

- EU applicant with Europe Association Agreement.

Defense and military agreements with the United States

- 1992, IMET.
- 1995, military assistance (Foreign Assistance Act).

Defense ties with the United States

Has hosted PfP exercises and been an active partner.

Russia

UN voting coincidence and freedom rating

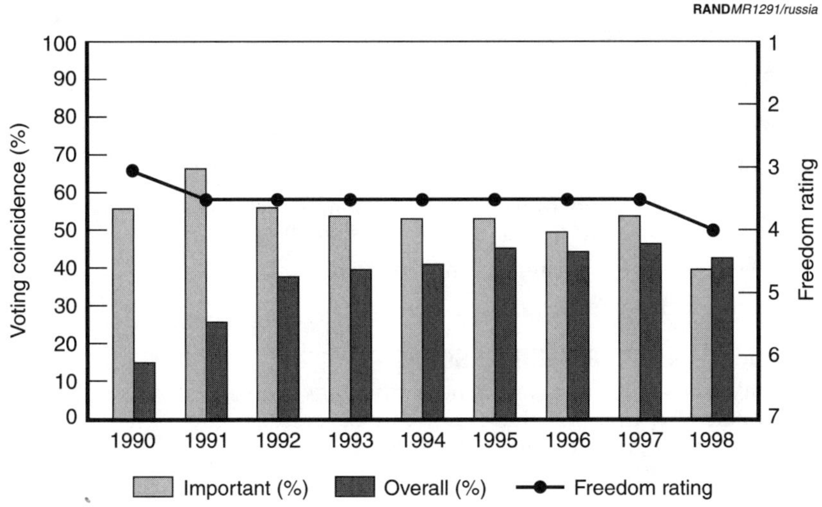

Participation in recent major operations

Desert Storm	Restore Hope	Uphold Democracy	Deliberate Force	Joint Endeavor	Allied Force	Joint Guardian
				✓		✓

Force size and defense spending

Ground force size	Total force size	Defense spending per soldier
420,000	1,240,000	approximately $20,000–$52,000

Alignment

Military:

- PfP and Founding Act with NATO.
- CIS (with military cooperation provisions).
- Increasing arms trade with China, India.
- Improving relations with Japan.
- Some defense trade with Iran.

- Moves toward improving ties with Laos.
- Range of arms sales and training arrangements worldwide.
- Military cooperation agreements and arrangements with CIS states, Bulgaria, Kuwait.

Nonmilitary:

- ASEAN consultative partner.
- ARF.

Defense and military agreements with the United States

- 1992, IMET
- 1992–present, cooperative threat reduction (nuclear, chemical, and biological weapons dismantlement assistance) agreements.
- 1996, nuclear test seismic monitoring stations.
- Arms control agreements

Defense ties with the United States

Russia's focus in PfP and NATO cooperation has been on contacts with the U.S. bilaterally. Contacts and talks at various levels continue. PfP involvement has not been high. Military relations with NATO greatly strained by Operation Allied Force (Kosovo).

Slovakia

UN voting coincidence and freedom rating

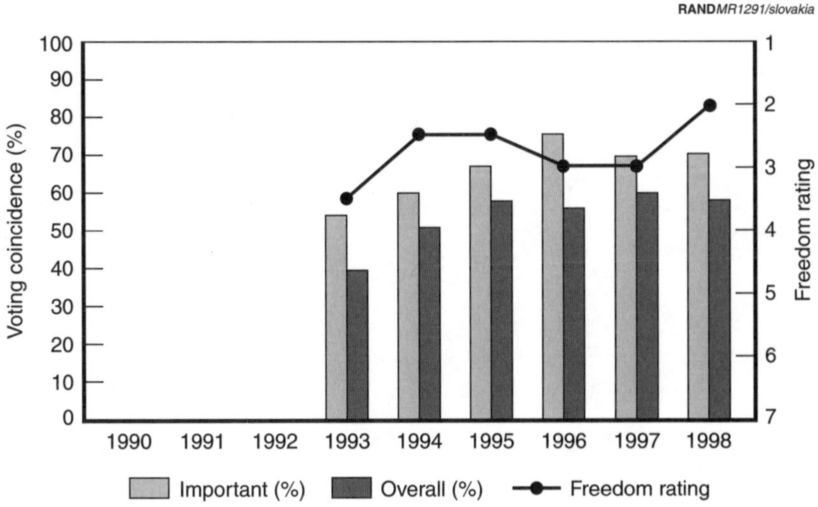

Participation in recent major operations

Desert Storm	Restore Hope	Uphold Democracy	Deliberate Force	Joint Endeavor	Allied Force	Joint Guardian
*						✓

*Czechoslovakia participated in Desert Storm with a small contingent, but its deployment is not counted here.

Force size and defense spending

Ground force size	Total force size	Defense spending per soldier
23,800	41,200	approximately $10,000–$11,000

Alignment

Military:

- PfP.
- WEU associate partner.

Nonmilitary:

- EU applicant with Europe Association Agreement.

Defense and military agreements with the United States

- 1995, protection of classified military information.

Defense ties with the United States

Has hosted PfP exercises.

Slovenia

UN voting coincidence and freedom rating

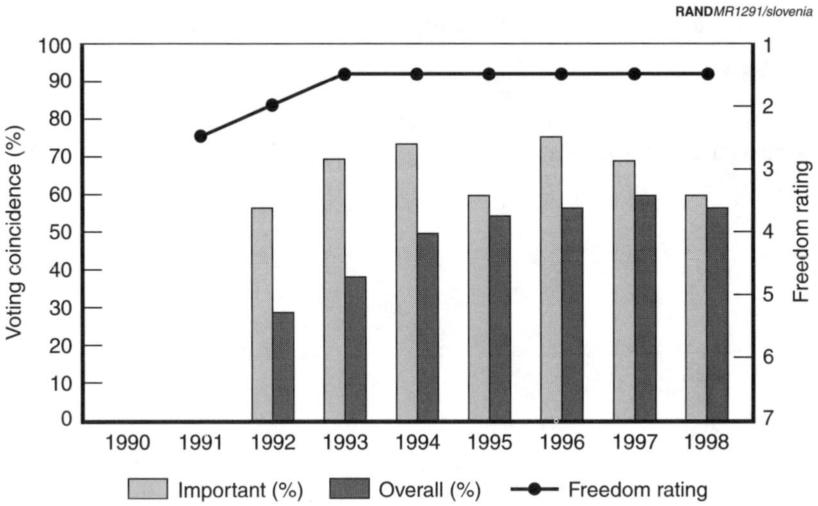

Participation in recent major operations

Desert Storm	Restore Hope	Uphold Democracy	Deliberate Force	Joint Endeavor	Allied Force	Joint Guardian

Force size and defense spending

Ground force size	Total force size	Defense spending per soldier
9,600	9,600	approximately $32,000

Alignment

Military:

- PfP.
- WEU associate partner.
- Supported by Italy for NATO membership.
- Slovenian-Hungarian-Italian Brigade.
- German training.

- Staff-level contacts with U.K.
- Military cooperation and training agreements with Austria and Hungary.

Nonmilitary:

- EU applicant with Europe Association Agreement.

Defense and military agreements with the United States

- 1993, IMET.

Defense ties with the United States

Active participant in PfP.

Spain

UN voting coincidence and freedom rating

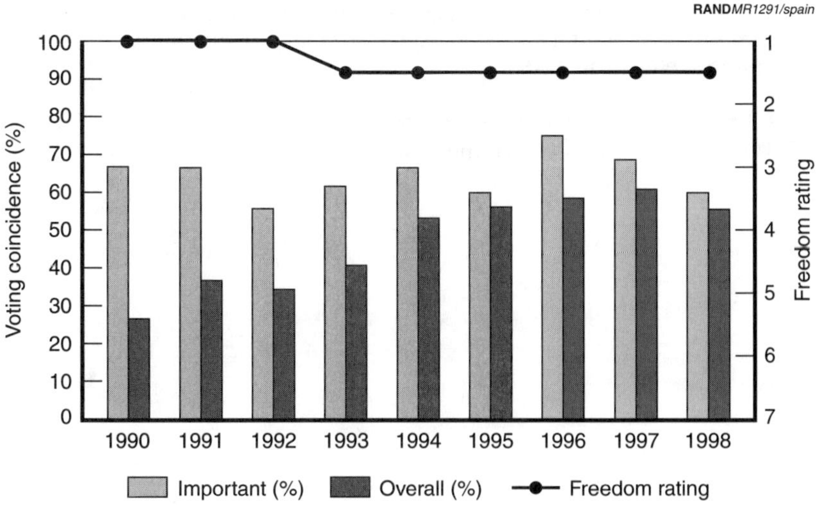

Participation in recent major operations

Desert Storm	Restore Hope	Uphold Democracy	Deliberate Force	Joint Endeavor	Allied Force	Joint Guardian
			✓	✓	✓	✓

Force size and defense spending

Ground force size	Total force size	Defense spending per soldier
128,500	197,500	approximately $39,000–$45,000

Alignment

Military:

- NATO.
- WEU (including European Corps).
- Joint Amphibious Brigade with Italy.
- Eurofor with France, Italy, Portugal.
- Some training in Chile.

Nonmilitary:

- EU.

Defense and military agreements with the United States

- 1953, mutual defense assistance.
- Range of agreements and arrangements on facilities assistance, offshore procurement, military assistance, data exchange, cooperation, logistic support, etc. through 1993.

Defense ties with the United States

NATO alliance and bilateral cooperation.

Sweden

UN voting coincidence and freedom rating

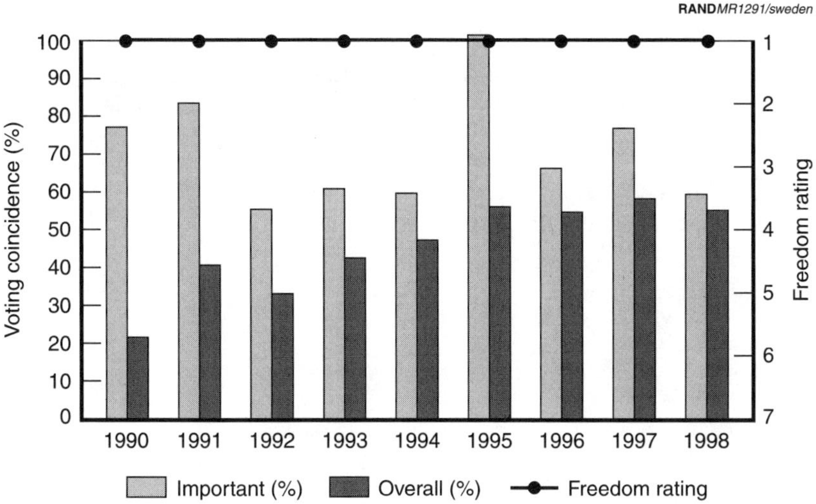

Participation in recent major operations

Desert Storm	Restore Hope	Uphold Democracy	Deliberate Force	Joint Endeavor	Allied Force	Joint Guardian
	✓			✓		✓

Force size and defense spending

Ground force size	Total force size	Defense spending per soldier
35,100	53,400	approximately $103,000–$116,000

Alignment

Military:

- PfP.
- WEU observer.
- NORDBAT (Nordic Battalion with Denmark and Norway).

Nonmilitary:

- EU.

Defense and military agreements with the United States

- 1952, weapons procurement.
- 1981, security of military information.
- 1987, defense procurement cooperation.
- 1992, scientist exchange.
- 1993, R&D.
- 1995, environmental protection cooperation.
- 1995, bilateral program in electromagnetic effects measurement and analysis.

Defense ties with the United States

PfP and associated humanitarian/peacekeeping.

Ukraine

UN voting coincidence and freedom rating

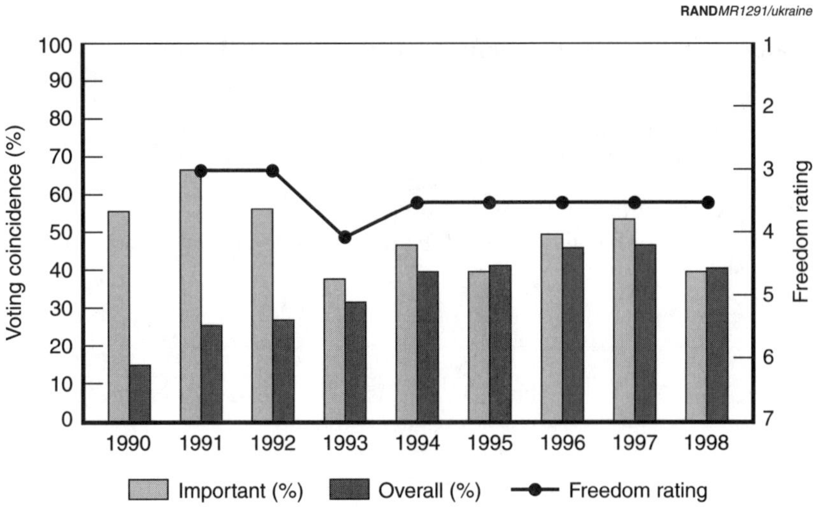

Participation in recent major operations

Desert Storm	Restore Hope	Uphold Democracy	Deliberate Force	Joint Endeavor	Allied Force	Joint Guardian
				✓		✓

Force size and defense spending

Ground force size	Total force size	Defense spending per soldier
161,500	387,400	approximately $3,000–$4,000

Alignment

Military:

- PfP.
- Distinctive Partnership with NATO.
- GUUAM.
- Seeking stronger Black Sea ties, as with Turkey.

- Range of agreements with Russia governing Black Sea Fleet and overall relations.
- Joint peacekeeping brigade with Poland.

Nonmilitary:

- CIS (without military cooperation provisions).

Defense and military agreements with the United States

- 1992, IMET
- 1993–present, cooperative threat reduction (nuclear, chemical, and biological weapons dismantlement assistance) agreements.

Defense ties with the United States

Active in PfP; strong bilateral and PfP program of ties, exercises, activities and visits.

United Kingdom

UN voting coincidence and freedom rating

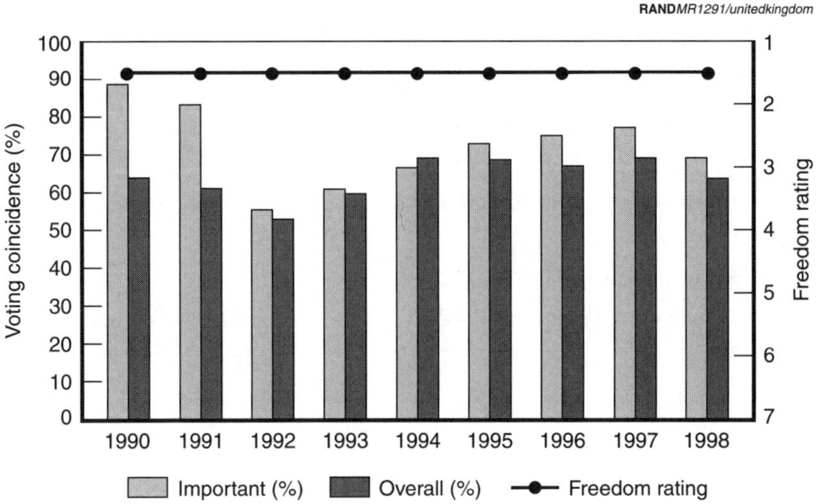

Participation in recent major operations

Desert Storm	Restore Hope	Uphold Democracy	Deliberate Force	Joint Endeavor	Allied Force	Joint Guardian
✓	✓		✓	✓	✓	✓

Force size and defense spending

Ground force size	Total force size	Defense spending per soldier
112,200	213,800	approximately $162,000–$167,000

Alignment

Military:

- Manila Pact (SEATO mutual defense agreement, still in force).
- NATO.
- WEU.
- Multinational Division (Central) with Belgium, Germany, Netherlands.
- European Air Corps with France, Germany, Italy.

- U.K.-Netherlands Amphibious Force.
- Five-Power Defense Arrangement.
- Cooperating with France and the U.S. to establish African Peacekeeping Force.
- Small presence in Norway.

Nonmilitary:

- EU, Commonwealth.

Defense and military agreements with the United States

- 1950, mutual defense assistance.
- Plethora of agreements on cooperation, facilities, sharing of information, etc.

Defense ties with the United States

Very close bilateral ties and contacts as well as NATO cooperation.

Central and South Asia (8 countries)

	Important Votes	Overall Votes	Freedom Rating	Number of Ops	Ground Strength	Per-Soldier Spending	Ties & Alignment
Bangladesh	0.34	0.26	3.17	2	101,000	$5,000	Low
India	0.24	0.15	3.33	2	980,000	$8,000–$11,000	Low
Kazakhstan	0.50	0.39	5.25	0	20,000	$14,000	Medium
Kyrgyzstan	0.33	0.26	4.00	0	9,800	$4,000	Low
Nepal	0.43	0.29	3.33	1	46,000	$800–$900	Low
Pakistan	0.33	0.24	4.28	2	520,000	$6,000	Medium
Turkmenistan	0.11	0.17	6.75	0	17,000	$6,000	Low
Uzbekistan	0.36	0.23	6.50	0	45,000	$6,000	Medium

NOTE: The first three columns report average values, 1990–1998.

KEY	Imp. Votes[a]	Overall Votes[a]	Freedom Rating	Number of Ops.	Ground Strength[b]	Per-Soldier Spending[c]	Alignment
High	0 to 33%	0 to 25%	1 to 3	4 to 7	> 60,000	>$94,000	Frequent regularized contacts, training, agreements, with U.S.
Medium	33 to 60%	25 to 45%	3.1 to 5.5	1 to 3	20,000–60,000	$19,000–$94,000	Infrequent or irregular contacts **or** strong ties with U.S. allies
Low	60 to 100%	45 to 100%	5.5 to 7	0	5,000–19,999	<$19,000	Few, sporadic, or no contacts

[a]For UN votes, the high, medium, and low range cutoffs represent the 0–25th, 25–75th, and 75–100th percentiles, respectively.

[b]Ground-strength low, medium, and high cutoffs are the assessed ability to contribute to multinational operations at a strength of approximately one brigade, one division, or more, respectively. Based on 1997 data.

[c]Defense-spending high, medium, and low cutoffs are the percentage of U.S. per-soldier spending at 20% or less, 20–50%, and over 50%, respectively. Based on 1997 data.

Bangladesh

UN voting coincidence and freedom rating

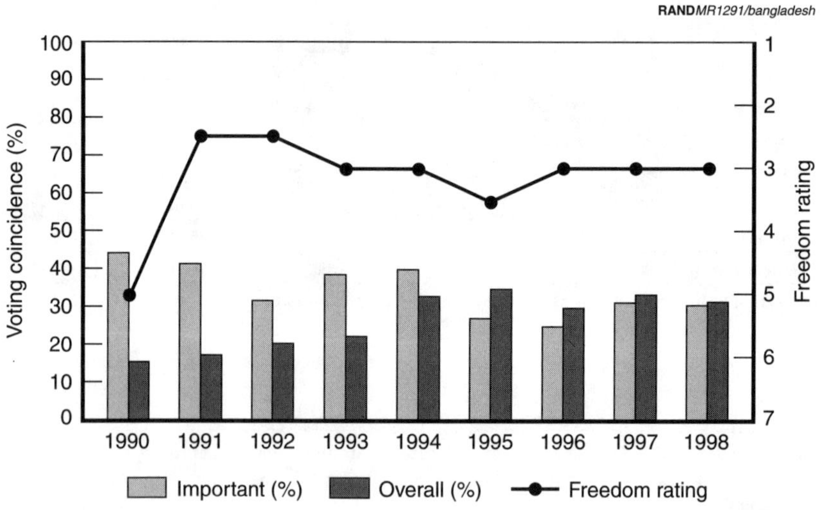

Participation in recent major operations

Desert Storm	Restore Hope	Uphold Democracy	Deliberate Force	Joint Endeavor	Allied Force	Joint Guardian
✓		✓				

Force size and defense spending

Ground force size	Total force size	Defense spending per soldier
101,000	121,000	approximately $5,000

Alignment

Nonmilitary:

- Commonwealth.

Defense and military agreements with the United States

- 1989, officer exchange.
- 1994, military assistance (Foreign Assistance Act).

India

UN voting coincidence and freedom rating

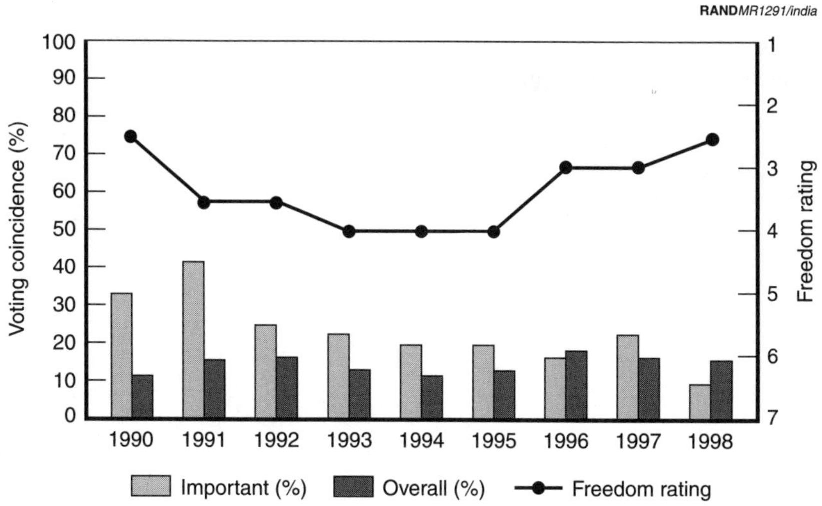

Participation in recent major operations

Desert Storm	Restore Hope	Uphold Democracy	Deliberate Force	Joint Endeavor	Allied Force	Joint Guardian
	✓	✓				

Force size and defense spending

Ground force size	Total force size	Defense spending per soldier
980,000	1,145,000	approximately $8,000–$11,000

Alignment

Military:

- Arms purchases from Russia.

Nonmilitary:

- Non-Aligned Movement.
- Commonwealth.
- ARF (ASEAN dialogue partner).

Defense and military agreements with the United States

- 1951, arms sales (with related understanding, 1958, and supplement, 1962).
- 1965, military assistance.
- 1995, defense agreement to increase bilateral defense cooperation.

Defense ties with the United States

Were improving prior to nuclear test.

Kazakhstan

UN voting coincidence and freedom rating

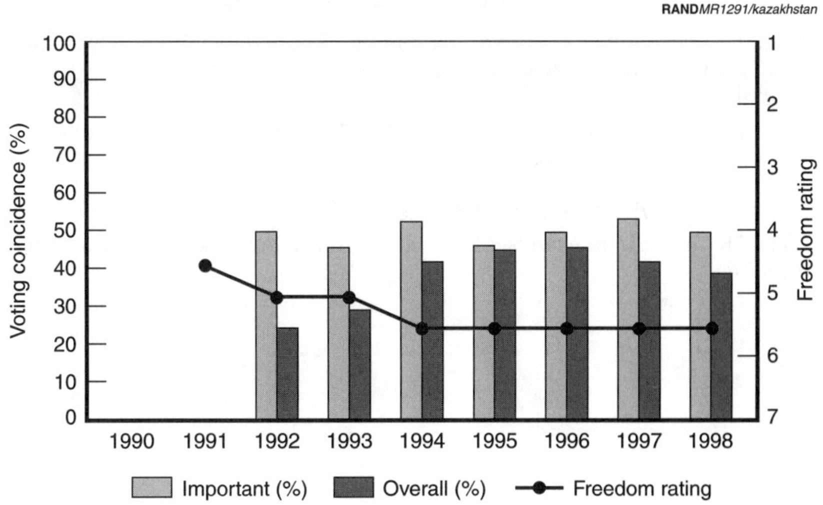

Participation in recent major operations

Desert Storm	Restore Hope	Uphold Democracy	Deliberate Force	Joint Endeavor	Allied Force	Joint Guardian

Force size and defense spending

Ground force size	Total force size	Defense spending per soldier
20,000	35,100	approximately $14,000

Alignment

Military:

- PfP.
- CIS (with military cooperation provisions).
- Training exchange with Russia.
- 16 defense-related agreements with Russia as of February 1996.
- Military cooperation agreement with Kyrgyzstan.

- Kazakh, Kyrgyz, and Uzbek joint committee with joint peacekeeping battalion.
- Agreement planned with Armenia.
- Recently, Kazakhstan is turning more toward West.

Defense and military agreements with the United States

- 1993–1995, with amendments extending—cooperative threat reduction (CTR: nuclear, chemical, and biological weapons dismantlement assistance).
- 1997, defense cooperation agreement (CTR, defense conversion, and assistance to the Kazakh military).

Defense ties with the United States

PfP, CTR, recent joint exercises focused on improving Kazakh peacekeeping training.

Kyrgyzstan

UN voting coincidence and freedom rating

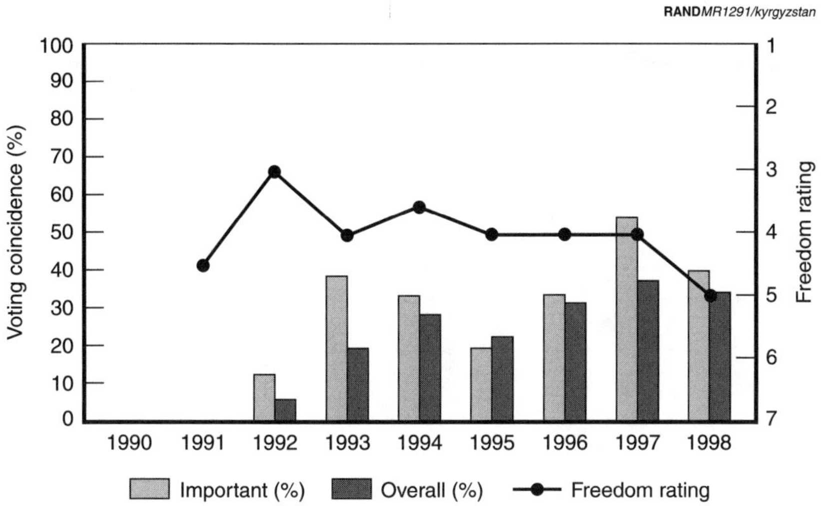

Participation in recent major operations

Desert Storm	Restore Hope	Uphold Democracy	Deliberate Force	Joint Endeavor	Allied Force	Joint Guardian

Force size and defense spending

Ground force size	Total force size	Defense spending per soldier
9,800	12,200	approximately $4,000

Alignment

Military:

- PfP.
- CIS (with military cooperation provisions).
- Training in Russia, Uzbekistan, Turkey.
- Kazakh, Kyrgyz, and Uzbek joint committee with joint peacekeeping battalion.
- Russian presence at space tracking posts, air defense, border troops.

- Cooperation with Danes agreed within PfP framework.
- MOD recently described Russia as primary partner for defense cooperation.

Defense and military agreements with the United States

- 1994, IMET.

Defense ties with the United States

PfP.

Nepal

UN voting coincidence and freedom rating

RANDMR1291/nepal

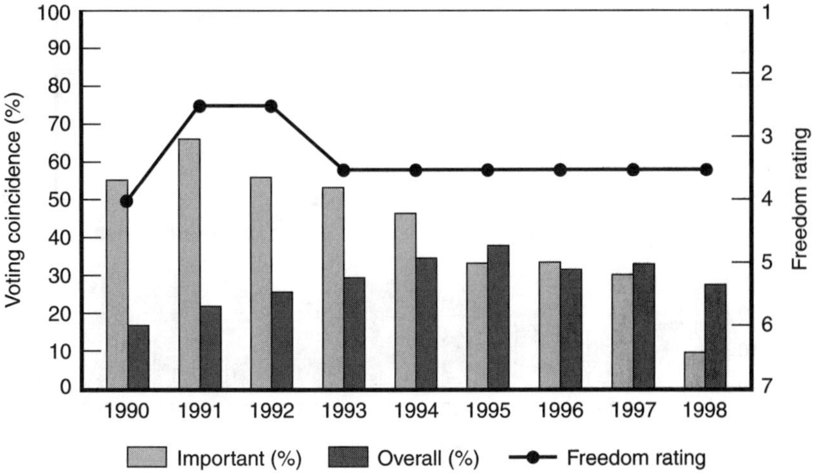

Participation in recent major operations

Desert Storm	Restore Hope	Uphold Democracy	Deliberate Force	Joint Endeavor	Allied Force	Joint Guardian
		✓				

Force size and defense spending

Ground force size	Total force size	Defense spending per soldier
46,000	46,000	approximately \$800–\$900

Alignment

Nonmilitary:

- Non-Aligned Movement.
- Good relations with India, China.

Defense and military agreements with the United States

- 1995, military assistance (Foreign Assistance Act).

Pakistan

UN voting coincidence and freedom rating

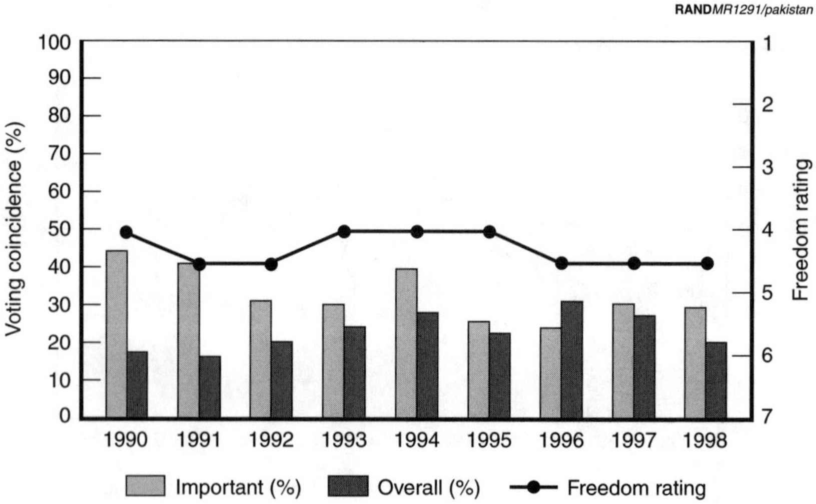

Participation in recent major operations

Desert Storm	Restore Hope	Uphold Democracy	Deliberate Force	Joint Endeavor	Allied Force	Joint Guardian
	✓	✓				

Force size and defense spending

Ground force size	Total force size	Defense spending per soldier
520,000	587,000	approximately $6,000

Alignment

Military:

- Close ties with China.
- Provides military personnel to strengthen gulf-state defenses.
- Some development of ties with Iran.
- Overtures to Ukraine.

Nonmilitary:

- Non-Aligned Movement.
- Commonwealth.

Defense and military agreements with the United States

- 1950, arms transfers.
- 1954, mutual defense assistance.
- 1955, defense support assistance (plus later assistance related agreements).
- 1959, cooperation.
- 1982, military information.
- 1986, IMET.
- 1990, coproduction of night vision equipment.

Defense ties with the United States

Prior to nuclear tests, relations were revitalizing with a bilateral consultative group.

Turkmenistan

UN voting coincidence and freedom rating

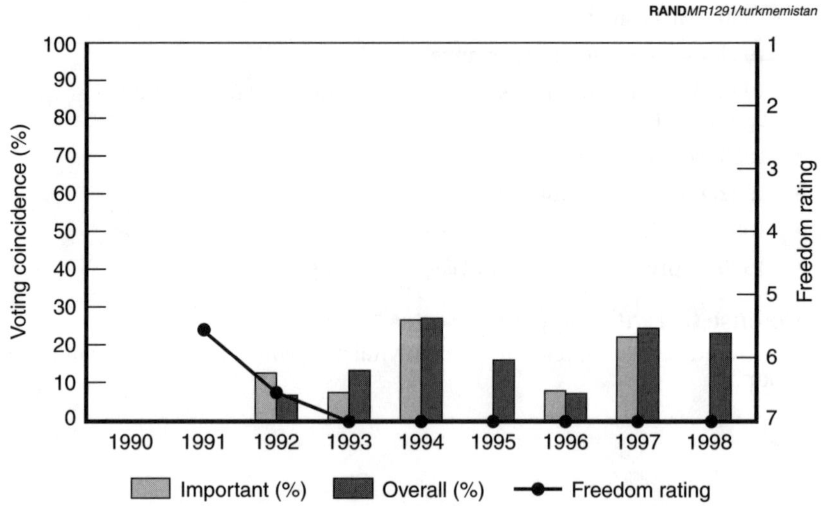

Participation in recent major operations

Desert Storm	Restore Hope	Uphold Democracy	Deliberate Force	Joint Endeavor	Allied Force	Joint Guardian

Force size and defense spending

Ground force size	Total force size	Defense spending per soldier
17,000	18,000	approximately $6,000

Alignment

Military:

- PfP.
- Close ties with Russia.
- Training in Russia and Ukraine.
- Close ties with Ukraine.
- Good relations with Saudi Arabia.

- Has been courted by Iran.
- Good relations with Turkey.

Nonmilitary:

- Non-Aligned Movement.
- CIS.

Defense and military agreements with the United States

- 1994, IMET.

Defense ties with the United States

PfP.

Uzbekistan

UN voting coincidence and freedom rating

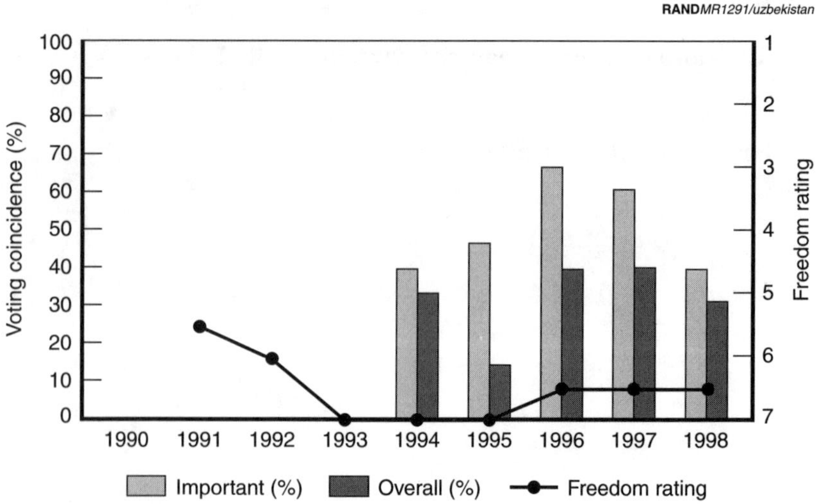

Participation in recent major operations

Desert Storm	Restore Hope	Uphold Democracy	Deliberate Force	Joint Endeavor	Allied Force	Joint Guardian

Force size and defense spending

Ground force size	Total force size	Defense spending per soldier
45,000	70,000	approximately $6,000

Alignment

Military:

- PfP.
- CIS (with military cooperation provisions).
- GUUAM.
- Kazakh, Kyrgyz, and Uzbek joint committee with joint peacekeeping battalion.
- Training in Russia.

- GUUAM membership, increasing ties with U.S. suggest turning away from Moscow.

Nonmilitary:

- Non-Aligned Movement.

Defense and military agreements with the United States

- 1995, IMET.
- 1997, nonproliferation and promotion of defense and military relations.

Defense ties with the United States

Increasing PfP, bilateral.

East Asia and Pacific (13 countries)

	Important Votes	Overall Votes	Freedom Rating	Number of Ops	Ground Strength	Per-Soldier Spending	Ties & Alignment
Australia	0.69	0.45	1.00	2	25,400	$147,000–$148,000	High
China	0.15	0.17	6.94	0	2,090,000	$3,000–$13,000	Low
Indonesia	0.33	0.23	5.89	0	220,000	$11,000–$17,000	Medium
Japan	0.66	0.45	1.56	0	147,700	$174,000–$230,000	High
South Korea	0.51	0.37	2.17	0	560,000	$22,000–$26,000	High
Laos	0.15	0.13	6.50	0	25,000	$2,000	Low
Malaysia	0.33	0.26	4.56	1	85,000	$22,000–$30,000	Medium
Mongolia	0.53	0.30	2.67	0		$2,000	Low
Myanmar	0.25	0.19	7.00	0	400,000	$5,000	Low
Philippines	0.34	0.26	2.94	0	70,000	$9,000–$13,000	Medium
Singapore	0.48	0.30	4.67	0	55,000	$59,000–$63,000	High
Thailand	0.39	0.27	3.44	0	150,000	$12,000–$14,000	High
Vietnam	0.12	0.13	7.00	0	420,000	$2,000	Low

NOTE: The first three columns report average values, 1990–1998.

KEY	Imp. Votes[a]	Overall Votes[a]	Freedom Rating	Number of Ops.	Ground Strength[b]	Per-Soldier Spending[c]	Alignment
High	0 to 33%	0 to 25%	1 to 3	4 to 7	> 60,000	>$94,000	Frequent regularized contacts, training, agreements, with U.S.
Medium	33 to 60%	25 to 45%	3.1 to 5.5	1 to 3	20,000–60,000	$19,000–$94,000	Infrequent or irregular contacts **or** strong ties with U.S. allies
Low	60 to 100%	45 to 100%	5.5 to 7	0	5,000–19,999	<$19,000	Few, sporadic, or no contacts

[a]For UN votes, the high, medium, and low range cutoffs represent the 0–25th, 25–75th, and 75–100th percentiles, respectively.

[b]Ground-strength low, medium, and high cutoffs are the assessed ability to contribute to multinational operations at a strength of approximately one brigade, one division, or more, respectively. Based on 1997 data.

[c]Defense-spending high, medium, and low cutoffs are the percentage of U.S. per-soldier spending at 20% or less, 20–50%, and over 50%, respectively. Based on 1997 data.

Australia

UN voting coincidence and freedom rating

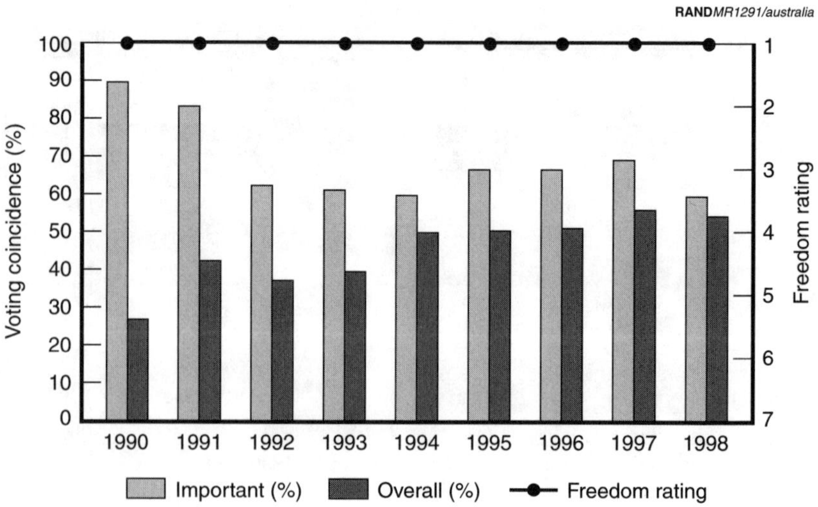

Participation in recent major operations

Desert Storm	Restore Hope	Uphold Democracy	Deliberate Force	Joint Endeavor	Allied Force	Joint Guardian
✓	✓					

Force size and defense spending

Ground force size	Total force size	Defense spending per soldier
25,400	57,400	approximately $147,000–$148,000

Alignment

Military:

- Manila Pact (SEATO mutual defense agreement, still in force).
- Five-Power Defense Arrangement.
- ANZUS.
- Increased military ties with Indonesia.

Nonmilitary:

- Commonwealth.
- ARF (ASEAN dialogue partner).

Defense and military agreements with the United States

- "Major non-NATO ally."
- 1951, mutual defense assistance (+1995 concerning "mutual defense commitments").
- 1960, weapons development.
- 1963, U.S. naval and 1969 joint space communications stations.
- 1976, 1984, 1992, training, exchanges.
- 1951–present, range of similar, related, and associated agreements and arrangements (including a 1966 joint space research agreement).

Defense ties with the United States

Approximately 700 U.S. personnel stationed at communications and intelligence locations in Australia. Regular exercise and cooperation program.

China

UN voting coincidence and freedom rating

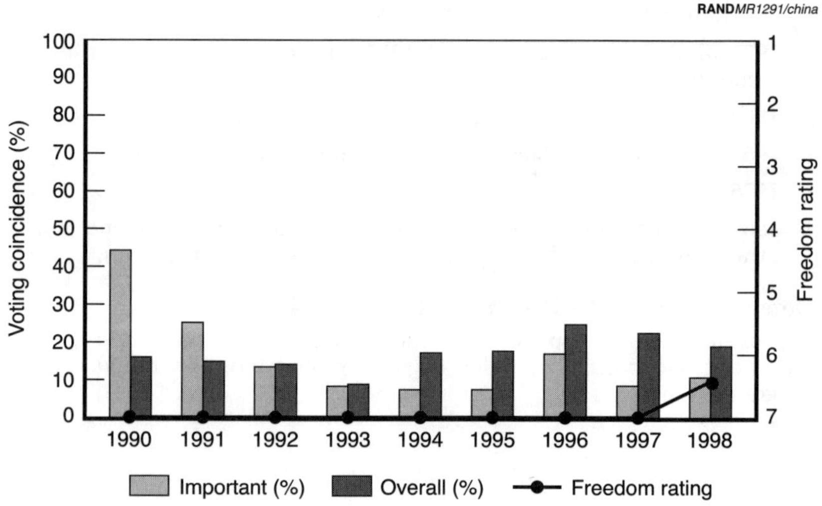

Participation in recent major operations

Desert Storm	Restore Hope	Uphold Democracy	Deliberate Force	Joint Endeavor	Allied Force	Joint Guardian

Force size and defense spending

Ground force size	Total force size	Defense spending per soldier
2,090,000	2,840,000	approximately $3,000–$13,000*

*PRC Defense spending estimates vary widely.

Alignment

Military:

- Improved and improving ties with Russia: "strategic partnership" announced during Yeltsin's 1997 visit.
- Relations with India improving.
- Long-standing good relations with Pakistan.

- Good relations with Bangladesh, which it views as in its sphere of influence.

Nonmilitary:

- ASEAN consultative partner.
- ARF.

Defense and military agreements with the United States

- 1998, maritime consultations.

Defense ties with the United States

High-level visits, exchanges. PLA has observed PfP exercise, participated in S&R exercise with U.S. Port visits to Hong Kong continue sporadically.

Indonesia

UN voting coincidence and freedom rating

RANDMR1291/indonesia

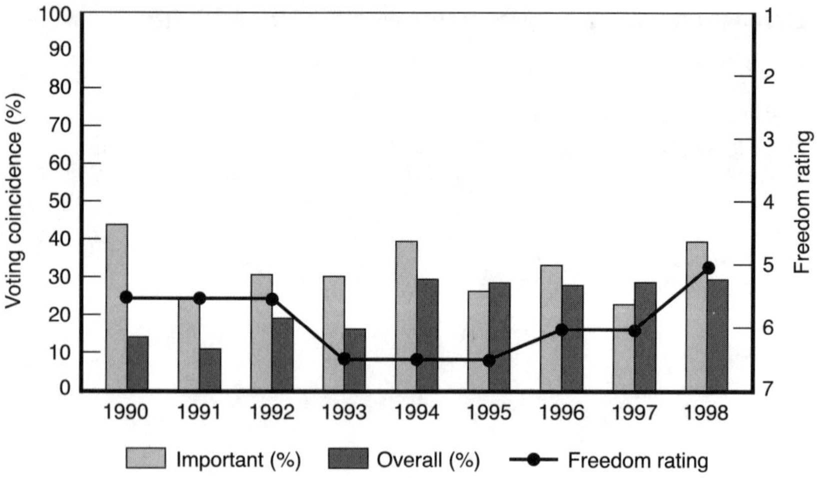

Participation in recent major operations

Desert Storm	Restore Hope	Uphold Democracy	Deliberate Force	Joint Endeavor	Allied Force	Joint Guardian

Force size and defense spending

Ground force size	Total force size	Defense spending per soldier
220,000	284,000	approximately $11,000–$17,000

Alignment

Military:

- Develops officer training in conjunction with U.S., Australia, U.K.

Nonmilitary:

- Non-Aligned Movement.
- ASEAN.
- ARF.

Defense and military agreements with the United States

- 1950, military assistance (constabulary equipment, with supplement changing it from grant to aid in 1953).
- 1958, weapons procurement.
- Other assistance agreements and arrangements through 1976, including assistance and training in 1976.

Defense ties with the United States

Military training, ship and aircraft visits, joint exercises, and mutual visits of ranking military officers.

Japan

UN voting coincidence and freedom rating

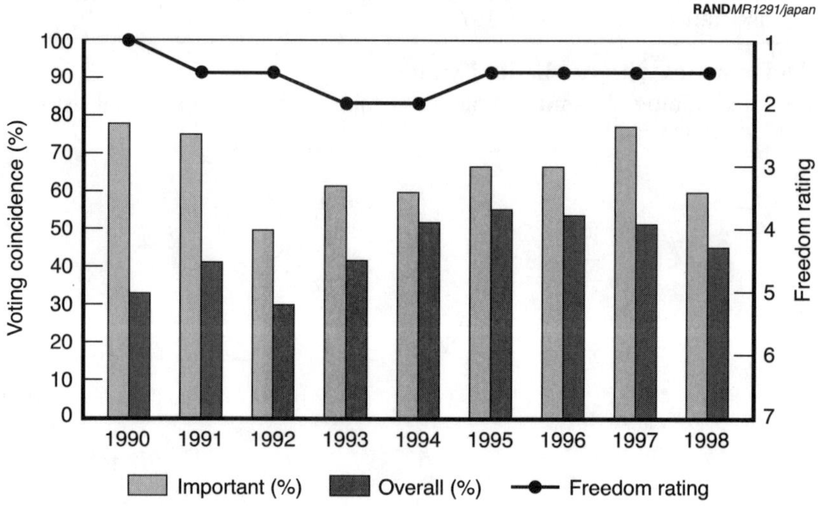

Participation in recent major operations

Desert Storm	Restore Hope	Uphold Democracy	Deliberate Force	Joint Endeavor	Allied Force	Joint Guardian

Force size and defense spending

Ground force size	Total force size	Defense spending per soldier
147,700	284,000	approximately $174,000–$230,000

Alignment

Military:

- Mutual defense commitment with U.S.
- Improved relations with China, Russia, South Korea.
- Expanded ties with the Middle East.
- Increasingly active in Africa and Latin America.

Nonmilitary:

- ARF (ASEAN dialogue partner).

Defense and military agreements with the United States

- "Major non-NATO ally."
- 1954, mutual defense assistance.
- Range of agreements through present covering military assistance, weapon procurement; weapons production, Japanese payment, etc.
- New 1997 accord on cooperation.

Defense ties with the United States

Very close ties, as befitting the alliance relationship. U.S. basing in-country; joint industrial development.

South Korea

UN voting coincidence and freedom rating

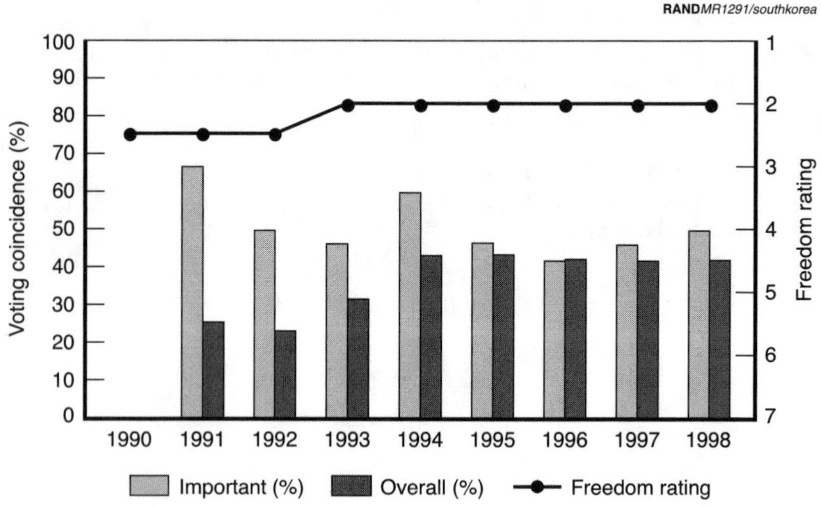

Participation in recent major operations

Desert Storm	Restore Hope	Uphold Democracy	Deliberate Force	Joint Endeavor	Allied Force	Joint Guardian

Force size and defense spending

Ground force size	Total force size	Defense spending per soldier
560,000	672,000	approximately $22,000–$26,000

Alignment

Military:

- Mutual defense treaty with U.S.
- Steady and significant improvement in relations with Japan.

Nonmilitary:

- ARF (ASEAN dialogue partner).

Defense and military agreements with the United States

- “Major non-NATO ally.”
- 1950, mutual defense assistance.
- 1953/54, mutual defense treaty.
- Range of cooperation; facilities; military assistance, SOFA, etc. agreements through present.

Defense ties with the United States

U.S. bases in South Korea. U.S. and South Korean forces exercise and prepare for conflict together. Some officer training in U.S.; domestic training along U.S. lines.

Laos

UN voting coincidence and freedom rating

RANDMR1291/laos

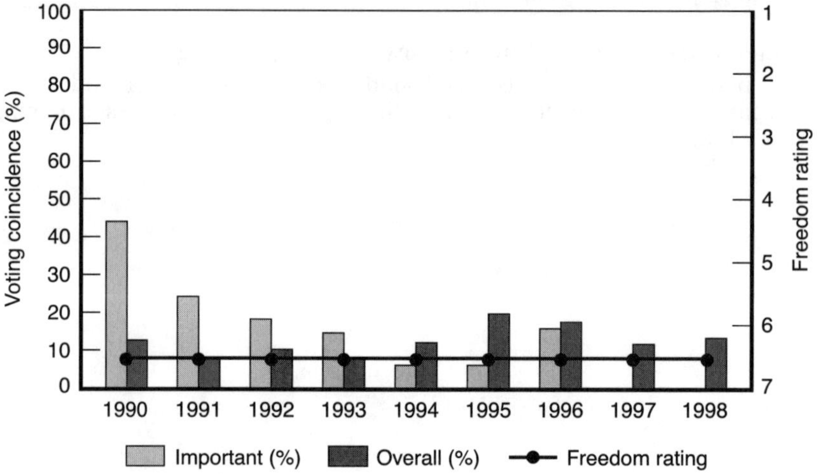

Participation in recent major operations

Desert Storm	Restore Hope	Uphold Democracy	Deliberate Force	Joint Endeavor	Allied Force	Joint Guardian

Force size and defense spending

Ground force size	Total force size	Defense spending per soldier
25,000	29,000	approximately $2,000

Alignment

Military:

- Working to improve relations with Thailand.
- Also improving with China, Australia, France, Japan, and India.
- Increasing military reliance on China.
- China may be providing training assistance, has three SIGINT stations in Laos.

- Some arms purchases from Russia.
- Military cooperation agreement signed with Russia in June 1997.

Nonmilitary:

- Non-Aligned Movement.
- ASEAN.
- ARF.

Defense and military agreements with the United States

- 1974, governing military assistance.
- 1993, military assistance (defense articles, training, and services).

Malaysia

UN voting coincidence and freedom rating

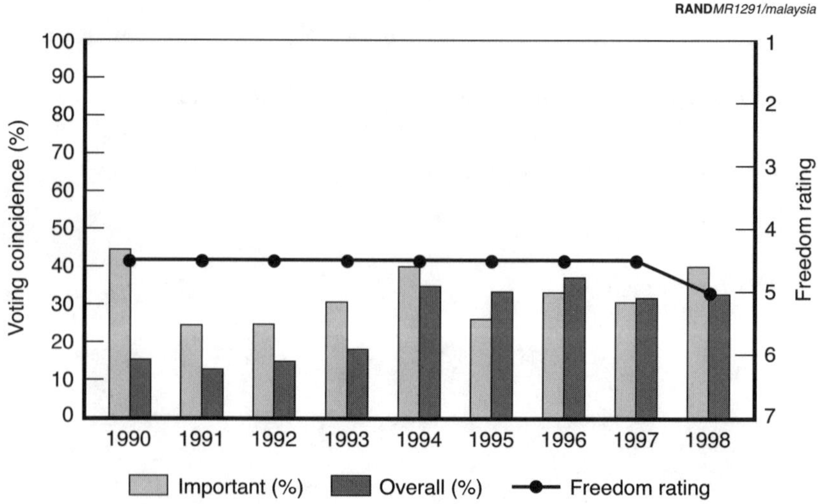

Participation in recent major operations

Desert Storm	Restore Hope	Uphold Democracy	Deliberate Force	Joint Endeavor	Allied Force	Joint Guardian
				✓		

Force size and defense spending

Ground force size	Total force size	Defense spending per soldier
85,000	111,500	approximately $22,000–$30,000

Alignment

Military:

- Five-Power Defense Arrangement (suspended October 1998; defense strategy under review).
- Small Australian presence.
- Recent exploration of improved ties with China.

Nonmilitary:

- Non-Aligned Movement.
- Commonwealth.
- ASEAN.
- ARF.

Defense and military agreements with the United States

- 1958, arms sales.
- 1972, military assistance-related.
- 1977, military assistance and training (International Security Assistance and Arms Export Control Act).
- 1991, military education exchange.
- 1994, acquisition and cross-servicing.

Defense ties with the United States

Exercises, ship visits.

Mongolia

UN voting coincidence and freedom rating

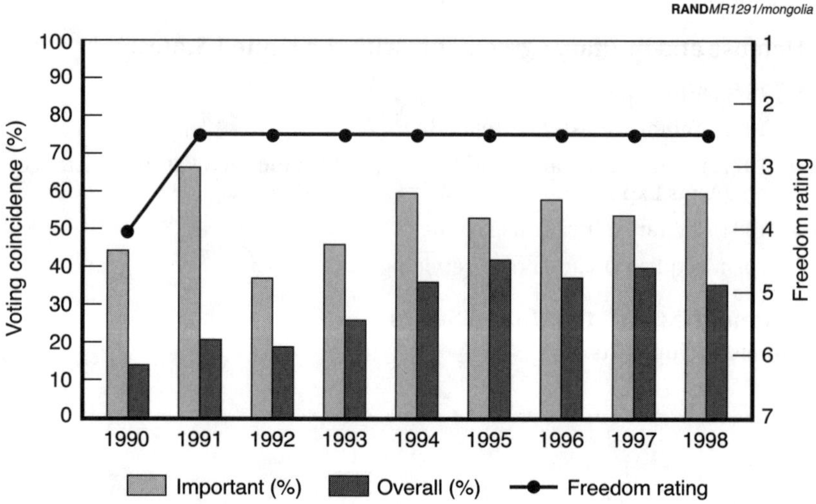

Participation in recent major operations

Desert Storm	Restore Hope	Uphold Democracy	Deliberate Force	Joint Endeavor	Allied Force	Joint Guardian

Force size and defense spending

Ground force size	Total force size	Defense spending per soldier
8,500	9,000	approximately $2,000

Alignment

Military:

- Expanding relations with Japan and South Korea.
- Issue of peacekeeping training raised with Canadians.
- Agreement for German training of Mongolian officers.
- May be considering joining PfP.
- 1991, Joint Declaration of Cooperation with Russia.

- 1993, Treaty of Friendship and Cooperation with Russia.
- Agreement with Germany to provide for Mongolian officer training in Germany.

Nonmilitary:

- Non-Aligned Movement.
- ARF.

Defense ties with the United States

U.S. has provided Army Corps of Engineers construction assistance to Mongolia.

Myanmar

UN voting coincidence and freedom rating

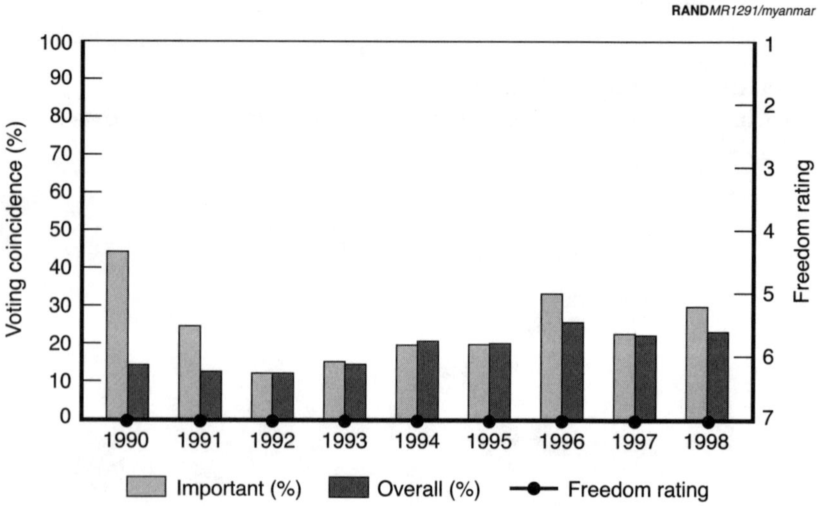

Participation in recent major operations

Desert Storm	Restore Hope	Uphold Democracy	Deliberate Force	Joint Endeavor	Allied Force	Joint Guardian

Force size and defense spending

Ground force size	Total force size	Defense spending per soldier
400,000	429,000	approximately $5,000

Alignment

Military:

- Closer ties to China: diplomatic support, aid, weapons.
- Relations with India improving somewhat, though India accepts refugees and voices support for democracy movement.

Nonmilitary:

- ASEAN.

- ARF.
- Non-Aligned Movement.

Defense and military agreements with the United States

- 1980, IMET.

Philippines

UN voting coincidence and freedom rating

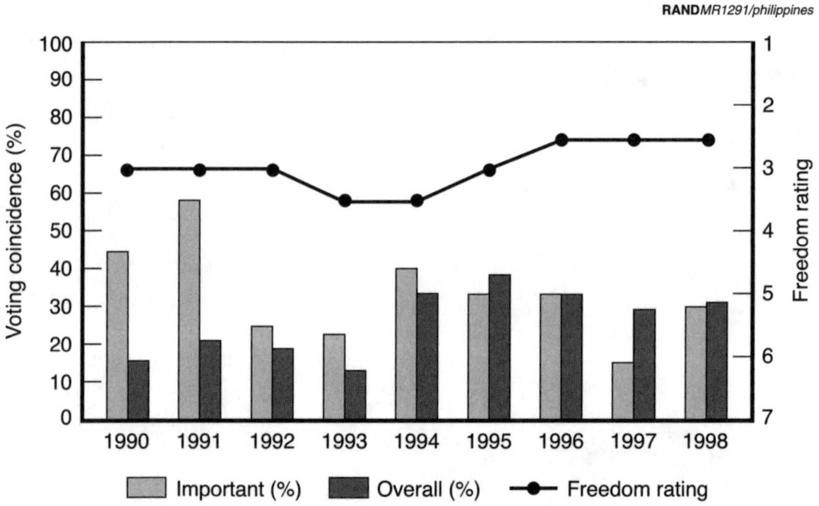

Participation in recent major operations

Desert Storm	Restore Hope	Uphold Democracy	Deliberate Force	Joint Endeavor	Allied Force	Joint Guardian

Force size and defense spending

Ground force size	Total force size	Defense spending per soldier
70,000	110,500	approximately $9,000–$13,000

Alignment

Military:

- Manila Pact (SEATO mutual defense agreement, still in force).
- Mutual defense commitment with U.S.
- Taiwan does some training in Philippines.

Nonmilitary:

- Non-Aligned Movement.

- ASEAN.
- ARF.

Defense and military agreements with the United States

- 1951, mutual defense.
- 1953, 1955, military assistance (amendments to 1955 agreement in 1956, 1957, 1958; related agreements 1979, 1980, 1981).
- Mutual defense board.
- Range of agreements on bases, SOFA, etc. through 1993; 1998 SOFA.

Defense ties with the United States

Recently agreed to resume port visits, joint exercises.

Singapore

UN voting coincidence and freedom rating

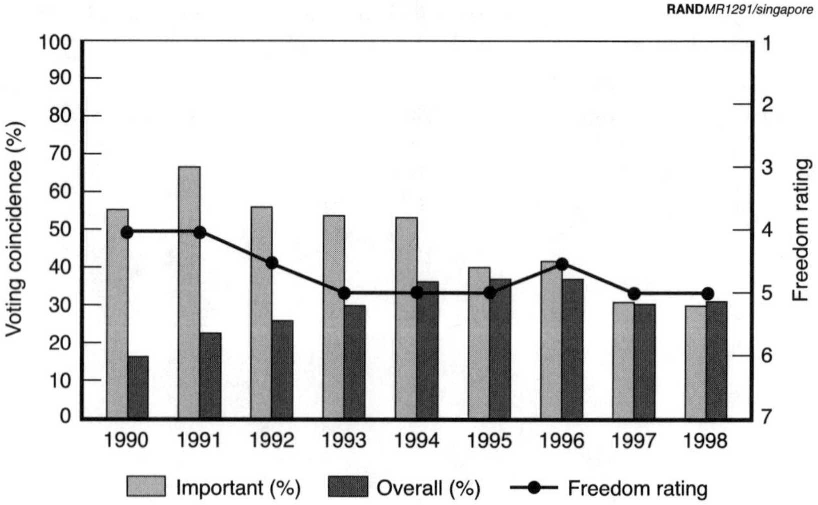

Participation in recent major operations

Desert Storm	Restore Hope	Uphold Democracy	Deliberate Force	Joint Endeavor	Allied Force	Joint Guardian

Force size and defense spending

Ground force size	Total force size	Defense spending per soldier
55,000	70,000	$59,000–$63,000

Alignment

Military:

- Five-Power Defense Arrangement.
- New Zealand has small presence (20 personnel).

Nonmilitary:

- Non-Aligned Movement.
- OAU.

- Commonwealth.
- ASEAN.
- ARF.

Defense and military agreements with the United States

- 1958, weapons procurement.
- 1977, training.
- 1981, 1982, personnel exchange.
- 1981, IMET.
- 1983, security of military information.
- 1991, C3 standards.
- 1993, Singapore in U.S. SOFA.
- 1995, R&D exchange.
- 1990, MOU on access.

Defense ties with the United States

Exercises, etc. Has offered increased base access; U.S. has small presence (150 personnel, Air Force and Navy).

Thailand

UN voting coincidence and freedom rating

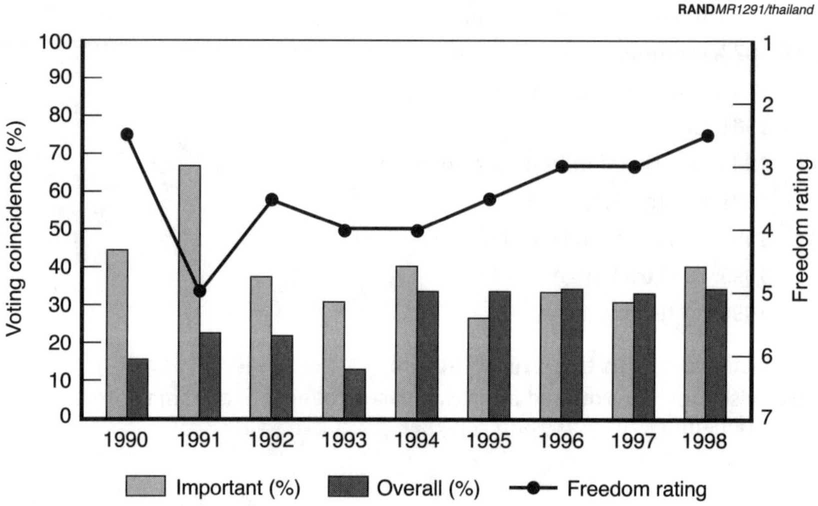

Participation in recent major operations

Desert Storm	Restore Hope	Uphold Democracy	Deliberate Force	Joint Endeavor	Allied Force	Joint Guardian

Force size and defense spending

Ground force size	Total force size	Defense spending per soldier
150,000	266,000	approximately $12,000–$14,000

Alignment

Military:

- Manila Pact (SEATO mutual defense agreement, still in force).
- Increasingly close ties with other ASEAN members: Indonesia, Malaysia, Philippines, Singapore, Brunei, Laos, Burma, and Vietnam.
- Some officers trained in Australia.
- Agreed to begin exercises with Vietnam.

- Australia, Singapore, and U.K. exercise and train in/with Thailand.

Nonmilitary:

- Non-Aligned Movement.
- ASEAN.
- ARF.

Defense and military agreements with the United States

- 1950, military assistance (related 1955, 1974/75).
- 1977, integrated communications; ammunition storage.
- 1983, security of military information.
- 1985, logistic support.
- 1986, IMET.
- 1987, war reserve stockpile program (amended, extended 1995).
- 1990, tactical C2.
- 1993, Thai Air Force air combat maneuvering instrumentation range facilities.

Defense ties with the United States

Regular and substantial joint exercise program; some officers trained in U.S..

Vietnam

UN voting coincidence and freedom rating

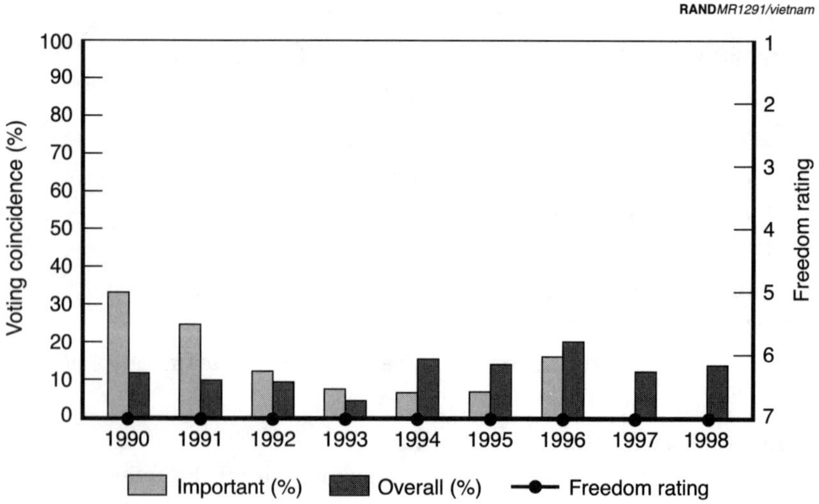

Participation in recent major operations

Desert Storm	Restore Hope	Uphold Democracy	Deliberate Force	Joint Endeavor	Allied Force	Joint Guardian

Force size and defense spending

Ground force size	Total force size	Defense spending per soldier
420,000	492,000	approximately $2,000

Alignment

Military:

- Reestablishing ties with Laos.
- Plan to begin conducting joint exercises with Thailand.

Nonmilitary:

- Non-Aligned Movement.
- ASEAN.

- ARF.

Defense and military agreements with the United States

- 1955 and 1972 (amended 1974), assistance-related agreements. Presumed defunct.

Defense ties with the United States

U.S.-Vietnamese joint task force on MIA.

Middle East (18 countries)

	Important Votes	Overall Votes	Freedom Rating	Number of Ops	Ground Strength	Per-Soldier Spending	Ties & Alignment
Armenia	0.44	0.36	3.93	0	58,600	$2,000	Low
Azerbaijan	0.39	0.30	5.44	0	53,300	$2,000	Medium
Bahrain	0.34	0.30	6.14	1	8,500	33,000–$34,000	High
Egypt	0.38	0.27	5.93	3	320,000	$6,000	High
Georgia	0.44	0.36	4.44	0	12,600	$3,000–$4,000	Medium
Israel	0.87	0.74	2.00	0	134,000	$49,000–$64,000	High
Jordan	0.30	0.26	3.93	1	90,000	$5,000–$6,000	High
Kuwait	0.50	0.36	4.93	2	11,000	$229,000–$236,000	High
Morocco	0.33	0.30	5.00	2	175,000	$7,000	High
Oman	0.32	0.29	5.93	1	25,000	$42,000	High
Qatar	0.35	0.30	6.50	1	8,500	$114,000	High
Saudi Arabia	0.42	0.29	7.00	2	70,000	$112,000–$113,000	High
Syria	0.16	0.14	7.00	1	215,000	$7,000	Low
Tunisia	0.34	0.29	5.50	1	27,000	$10,000–$11,000	Medium
Turkey	0.54	0.48	4.36	5	525,000	$12,000–$13,000	High
UAE	0.36	0.31	5.57	3	59,000	$38,000	High
Yemen	0.24	0.24	5.29	0	61,000	$6,000	Low

NOTE: The first three columns report average values, 1990–1998.

KEY	Imp. Votes[a]	Overall Votes[a]	Freedom Rating	Number of Ops.	Ground Strength[b]	Per-Soldier Spending[c]	Alignment
High	0 to 33%	0 to 25%	1 to 3	4 to 7	> 60,000	>$94,000	Frequent regularized contacts, training, agreements, with U.S.
Medium	33 to 60%	25 to 45%	3.1 to 5.5	1 to 3	20,000–60,000	$19,000–$94,000	Infrequent or irregular contacts or strong ties with U.S. allies
Low	60 to 100%	45 to 100%	5.5 to 7	0	5,000–19,999	<$19,000	Few, sporadic, or no contacts

[a]For UN votes, the high, medium, and low range cutoffs represent the 0–25th, 25–75th, and 75–100th percentiles, respectively.

[b]Ground-strength low, medium, and high cutoffs are the assessed ability to contribute to multinational operations at a strength of approximately one brigade, one division, or more, respectively. Based on 1997 data.

[c]Defense-spending high, medium, and low cutoffs are the percentage of U.S. per-soldier spending at 20% or less, 20–50%, and over 50%, respectively. Based on 1997 data.

Armenia

UN voting coincidence and freedom rating

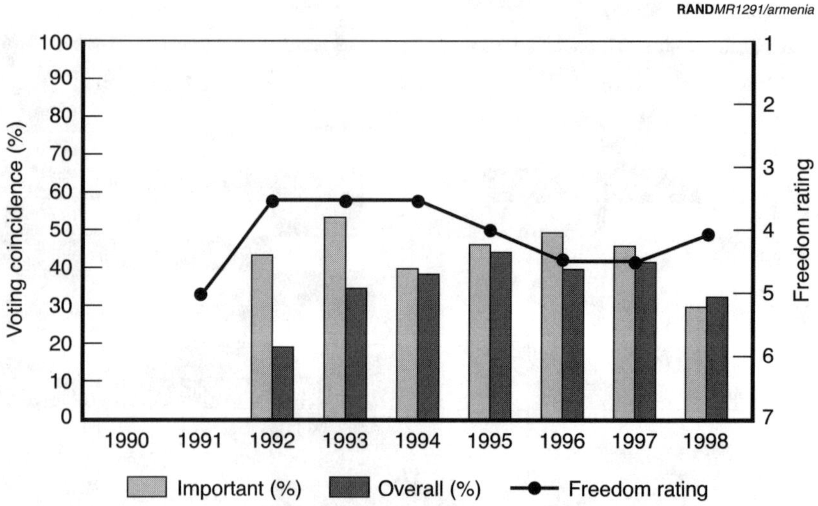

Participation in recent major operations

Desert Storm	Restore Hope	Uphold Democracy	Deliberate Force	Joint Endeavor	Allied Force	Joint Guardian

Force size and defense spending

Ground force size	Total force size	Defense spending per soldier
58,600	60,000	approximately $2,000

Alignment

Military:

- PfP.
- CIS (with military cooperation provisions).
- Close ties with Russia.
- Russian troop presence.

- Considerable training elsewhere in FSU (senior officer training in 2- to 3-year courses in Belarus and Russia).
- Participated in first PfP exercise in October 1997, with personnel from Greece, Netherlands, Bulgaria, and Albania (held in Greece).

Azerbaijan

UN voting coincidence and freedom rating

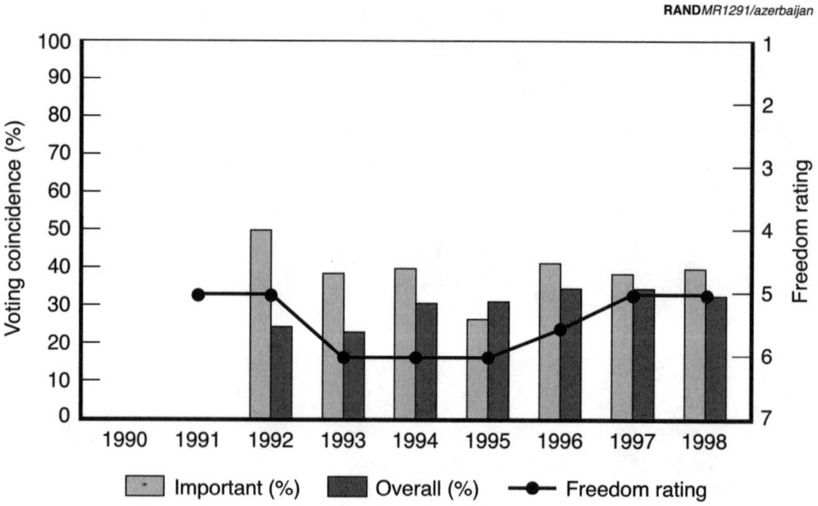

Participation in recent major operations

Desert Storm	Restore Hope	Uphold Democracy	Deliberate Force	Joint Endeavor	Allied Force	Joint Guardian

Force size and defense spending

Ground force size	Total force size	Defense spending per soldier
53,300	66,700	approximately $2,000

Alignment:

Military:

- PfP.
- GUUAM.
- Turkish officers on contract provide some training.
- Azeri forces will serve in KFOR under Turkish command.

Nonmilitary:

- CIS.

Defense ties with the United States

PfP, has offered base access to U.S.

Bahrain

UN voting coincidence and freedom rating

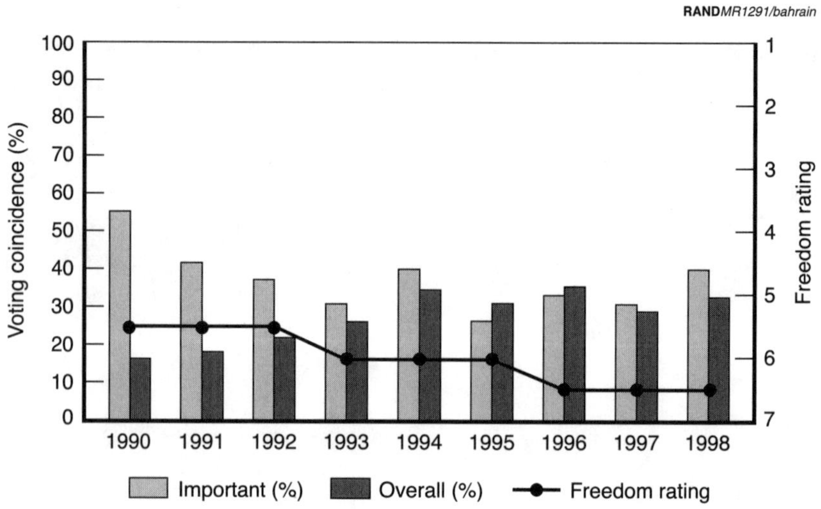

Participation in recent major operations

Desert Storm	Restore Hope	Uphold Democracy	Deliberate Force	Joint Endeavor	Allied Force	Joint Guardian
✓						

Force size and defense spending

Ground force size	Total force size	Defense spending per soldier
8,500	11,000	approximately $33,000–$34,000

Alignment

Military:

- Arab League.
- Gulf Cooperation Council.
- Training with other GCC states, including Saudi.

Nonmilitary:

- Non-Aligned Movement.

Defense and military agreements with the United States

- 1977, limited SOFA (admin support unit).
- 1992, military assistance (Foreign Assistance Act +), prepositioning, force access, and military training.
- 1994, cross-servicing.

Defense ties with the United States

U.S. forces deployed for Operation Southern Watch.

Egypt

UN voting coincidence and freedom rating

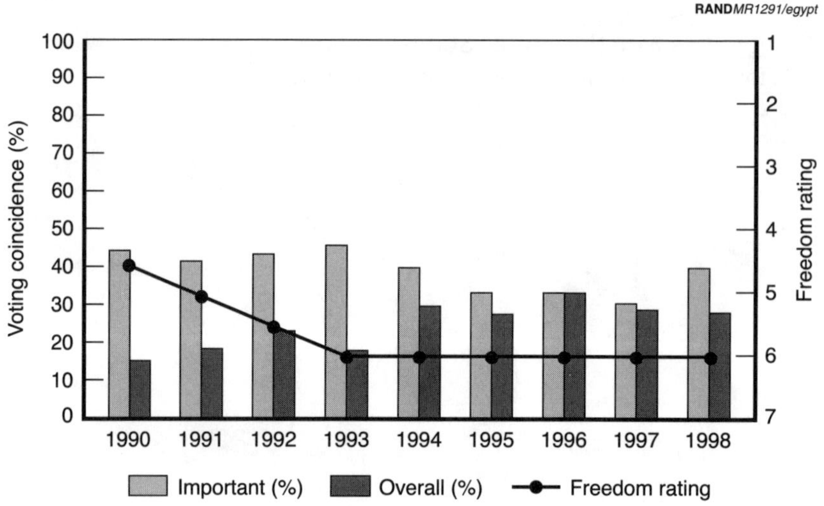

Participation in recent major operations

Desert Storm	Restore Hope	Uphold Democracy	Deliberate Force	Joint Endeavor	Allied Force	Joint Guardian
✓	✓			✓		

Force size and defense spending

Ground force size	Total force size	Defense spending per soldier
320,000	450,000	approximately $6,000

Alignment

Military:

- Arab League.
- Extensive exercises with the U.K., focusing on integrating air and maritime operations, mobile warfare, and special forces operations.

Nonmilitary:

- Non-Aligned Movement; OAU.

Defense and military agreements with the United States

- "Major non-NATO ally."
- 1952, agreement relating to mutual defense assistance (amended 1952).
- 1981, SOFA.
- 1982, security of military information.
- 1988, technical exchange.
- 1988, M1A1 coproduction (amended 1989).
- 1991, cooperative computer development.

Defense ties with the United States

Combined military exercises: "Bright Star" series with U.S. and U.K. + France, Italy, Kuwait, UAE (expanded from bilateral in 1995 to include Gulf War coalition members). Other extensive exercises with U.S. Primarily U.S. doctrine. U.S. deployments to Egypt.

Georgia

UN voting coincidence and freedom rating

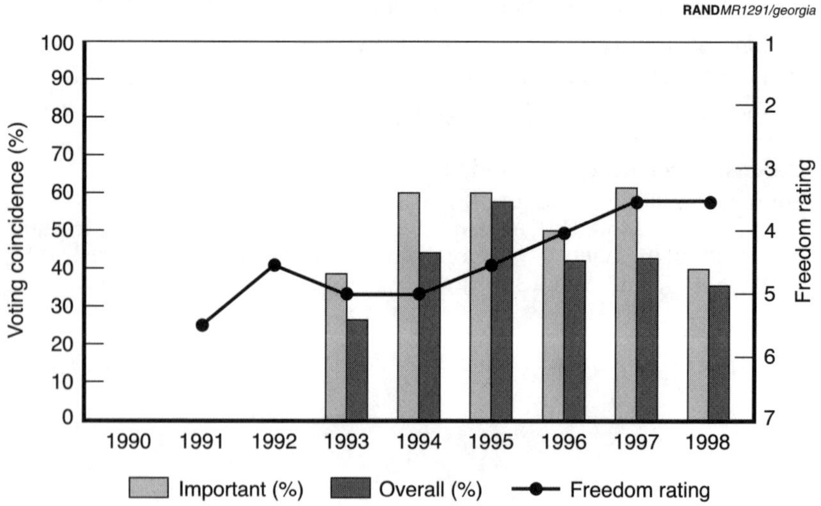

Participation in recent major operations

Desert Storm	Restore Hope	Uphold Democracy	Deliberate Force	Joint Endeavor	Allied Force	Joint Guardian

Force size and defense spending

Ground force size	Total force size	Defense spending per soldier
12,600	33,200	approximately $3,000–$4,000

Alignment

Military:

- PfP.
- GUUAM.
- Cadets study in Russian military academies.
- 1995, treaty with Russia on closer military cooperation.
- 1997, military cooperation pact with Turkey.

- Turks to train army officers.
- Turkey also providing assistance to Defense Ministry and border guards.
- Participated in Russian-led military operations.
- Georgian forces will serve in Kosovo under Turkish command.
- Russian troop presence.

Nonmilitary:

- Partnership and cooperation agreement with the EU; CIS.

Defense and military agreements with the United States

- 1993, IMET.
- 1997, cooperation in nonproliferation and promotion of defense and military relations.

Defense ties with the United States

PfP; U.S. participation in Georgia-hosted ISO PfP exercise in June 1997. Increasing military relationship.

Israel

UN voting coincidence and freedom rating

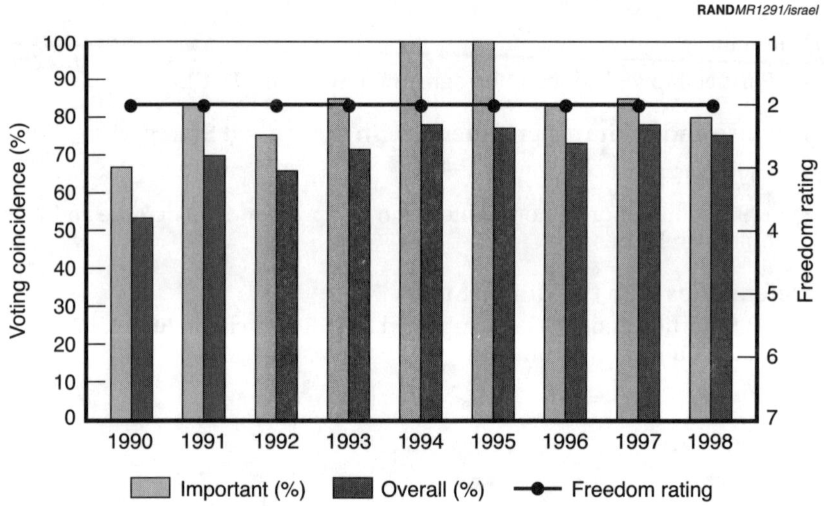

Participation in recent major operations

Desert Storm	Restore Hope	Uphold Democracy	Deliberate Force	Joint Endeavor	Allied Force	Joint Guardian

Force size and defense spending

Ground force size	Total force size	Defense spending per soldier
134,000	175,000	approximately $49,000–$64,000

Alignment

Military:

- Some training in Chile.
- Normalized relations with Egypt, Jordan.
- Recently, close military contacts fostered with Turkey.

Defense and military agreements with the United States

- "Major non-NATO ally."

- 1952, agreement relating to mutual defense assistance.
- 1965, weapons procurement (two, one for general procurement, the other for purchase of Israeli goods).
- 1991, (EIF 1994) reciprocal SOFAs.
- Range of logistics, production, equipment, facilities, etc. agreements through present.

Defense ties with the United States

Joint Political Military Group, meets twice a year. Joint military planning and combined exercises. Has collaborated on military research and weapons development.

Jordan

UN voting coincidence and freedom rating

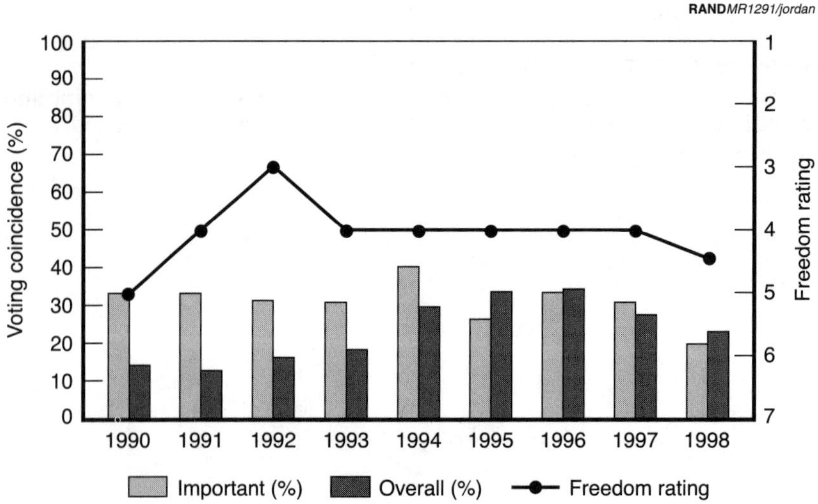

Participation in recent major operations

Desert Storm	Restore Hope	Uphold Democracy	Deliberate Force	Joint Endeavor	Allied Force	Joint Guardian
				✓		

Force size and defense spending

Ground force size	Total force size	Defense spending per soldier
90,000	104,100	approximately $5,000–$6,000

Alignment

Military:

- Arab League.
- Military exercise with Turkey in April 1998 reflects increase in ties.

Nonmilitary:

- Non-Aligned Movement.

Defense and military agreements with the United States

- "Major non-NATO ally."
- 1974, 1976, 1979 (and 1982 amendment), 1980 (1982 amendment) agreements governing military assistance (defense articles and services).
- 1988, mutual support.

Defense ties with the United States

Denied support in Gulf War. Since then, support for peace process has helped to improve relations. Ties include high-level visits, etc.

Kuwait

UN voting coincidence and freedom rating

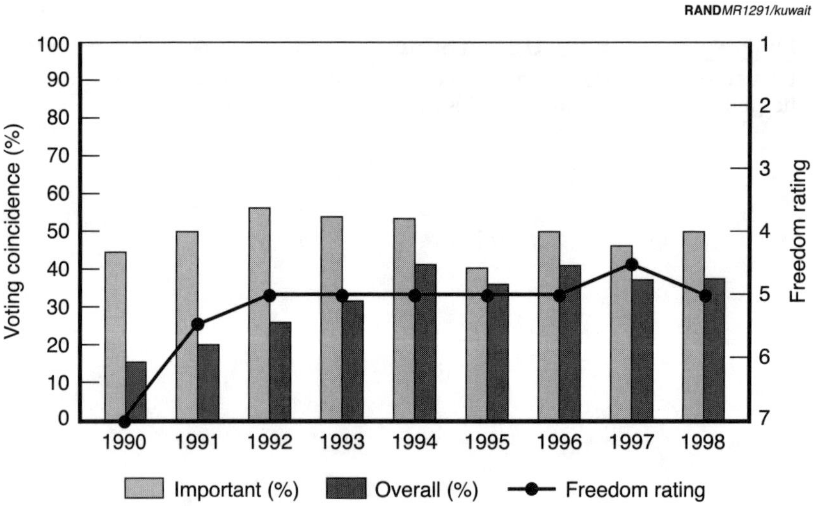

Participation in recent major operations

Desert Storm	Restore Hope	Uphold Democracy	Deliberate Force	Joint Endeavor	Allied Force	Joint Guardian
✓	✓					

Force size and defense spending

Ground force size	Total force size	Defense spending per soldier
11,000	15,300	approximately $229,000–$236,000

Alignment

Military:

- Arab League.
- Gulf Cooperation Council.
- Close ties with Egypt and Syria.
- Joint exercises with U.S., U.K., French, other forces.

Nonmilitary:

- Non-Aligned Movement.

Defense and military agreements with the United States

- 1975, Kuwaiti procurement of defense articles and services (arms sales).
- 1976, technical security.
- 1991/92, agreement covering military access, exercises, joint training, and weapons stocks.

Defense ties with the United States

Joint exercises; U.S. presence.

Morocco

UN voting coincidence and freedom rating

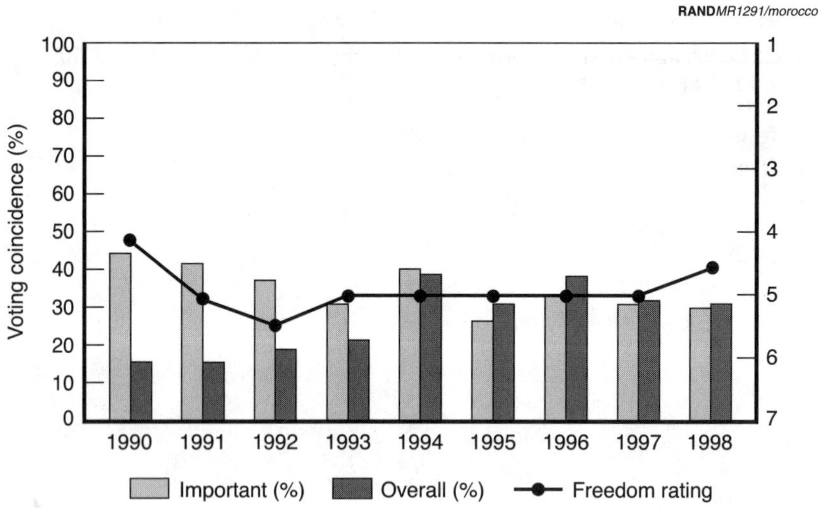

Participation in recent major operations

Desert Storm	Restore Hope	Uphold Democracy	Deliberate Force	Joint Endeavor	Allied Force	Joint Guardian
	✓			✓		

Force size and defense spending

Ground force size	Total force size	Defense spending per soldier
175,000	196,300	approximately $7,000

Alignment

Military:

- Arab League.
- Close relations with Saudi Arabia and the Persian Gulf states.
- Training in France.
- Spain maintains some presence in Spanish-owned enclaves.

Nonmilitary:

- Non-Aligned Movement.
- Arab Maghreb Union.

Defense and military agreements with the United States

- Treaty of Peace and Friendship from 1787 still in force.
- 1982, use of facilities.

Defense ties with the United States

Rights of transit through its airfields for U.S. forces. Joint exercises with various U.S. armed forces. U.S. advisers.

Oman

UN voting coincidence and freedom rating

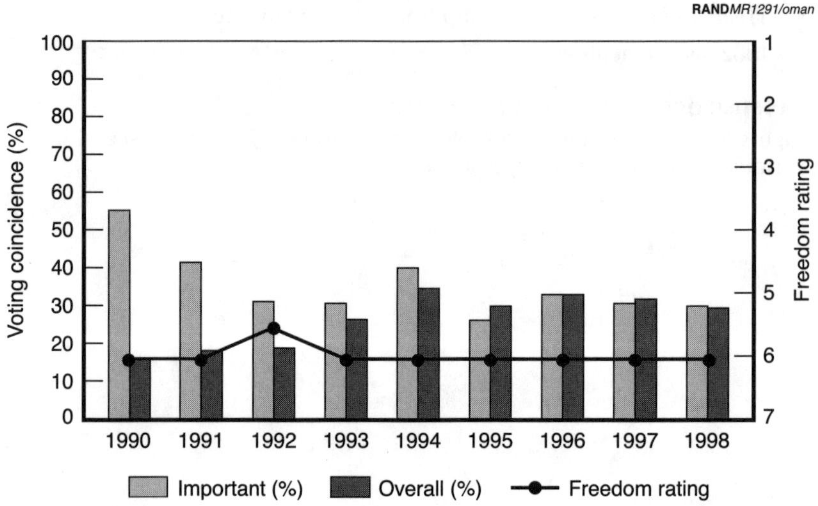

Participation in recent major operations

Desert Storm	Restore Hope	Uphold Democracy	Deliberate Force	Joint Endeavor	Allied Force	Joint Guardian
✓						

Force size and defense spending

Ground force size	Total force size	Defense spending per soldier
25,000	43,500	approximately $42,000

Alignment

Military:

- Gulf Cooperation Council.
- Arab League; friendly ties with U.K., France.
- Some ties to Iran.
- Outreach to Central Asia.
- Considerable training in U.K. and U.S.

Nonmilitary:

- Non-Aligned Movement.

Defense and military agreements with the United States

- 1986, IMET.
- 1992, military assistance (grants under Foreign Assistance Act + articles, training, and services).
- 1980, access to military facilities.
- 1986, prepositioning (1987 related agreement).

Defense ties with the United States

Training, prepositioning.

Qatar

UN voting coincidence and freedom rating

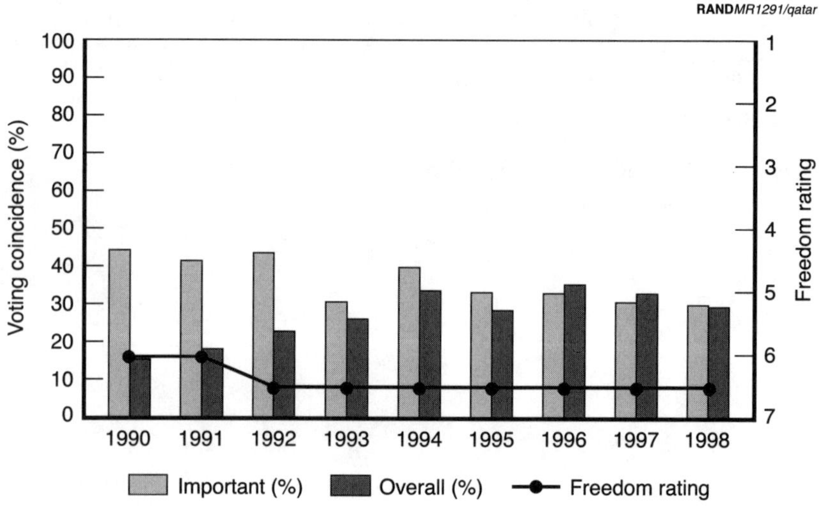

Participation in recent major operations

Desert Storm	Restore Hope	Uphold Democracy	Deliberate Force	Joint Endeavor	Allied Force	Joint Guardian
✓						

Force size and defense spending

Ground force size	Total force size	Defense spending per soldier
8,500	11,800	approximately $114,000

Alignment

Military:

- Gulf Cooperation Council.
- Arab League.
- Some French training.
- Egyptian advisers in past.

Nonmilitary:

- Non-Aligned Movement.

Defense and military agreements with the United States

- 1992, bilateral defense cooperation agreement provides for U.S. access to Qatari bases, prepositioning of U.S. materiel, and future combined military exercises.

Defense ties with the United States

Close. Prepositioning, joint exercises.

Saudi Arabia

UN voting coincidence and freedom rating

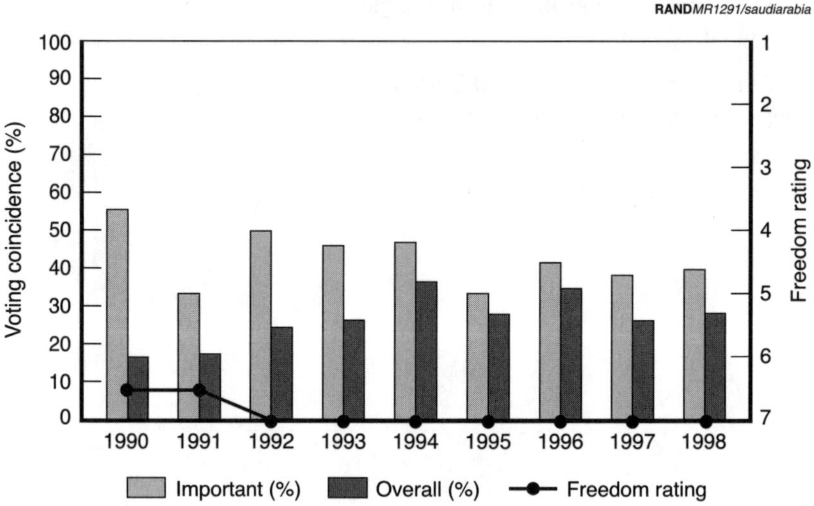

Participation in recent major operations

Desert Storm	Restore Hope	Uphold Democracy	Deliberate Force	Joint Endeavor	Allied Force	Joint Guardian
✓	✓					

Force size and defense spending

Ground force size	Total force size	Defense spending per soldier
70,000	162,500	approximately $112,000–$113,000

Alignment

Military:

- Gulf Cooperation Council.
- Arab League.
- French, U.S., and U.K. presence.
- Training in France, U.K., U.S.

Nonmilitary:

- Non-Aligned Movement.

Defense and military agreements with the United States

- 1960s-era agreements on transfer of military supplies and equipment, construction of military facilities; weapons procurement; SOFA for F-5 maintenance and operation.
- 1950s, 1960s, 1970s military assistance, training agreements.

Defense ties with the United States

Close informal ties; U.S. presence, joint exercises. Some training in U.S.

Syria

UN voting coincidence and freedom rating

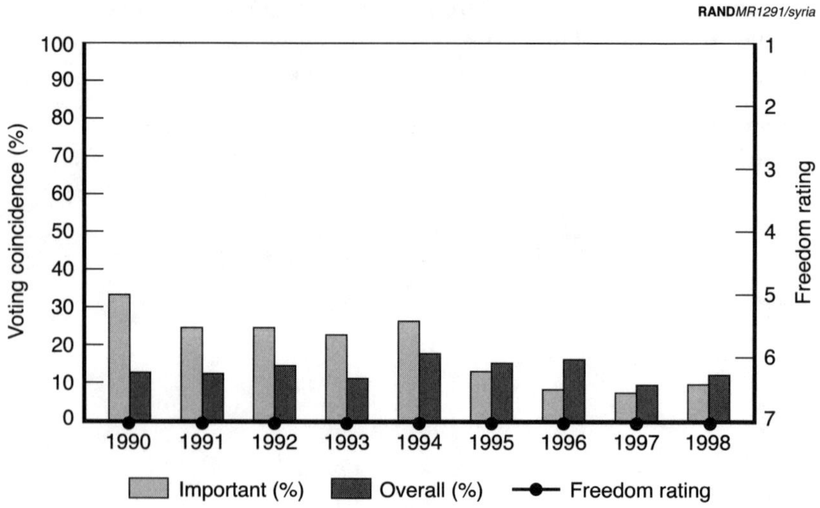

Participation in recent major operations

Desert Storm	Restore Hope	Uphold Democracy	Deliberate Force	Joint Endeavor	Allied Force	Joint Guardian
✓						

Force size and defense spending

Ground force size	Total force size	Defense spending per soldier
215,000	320,000	approximately $7,000

Alignment

Military:

- Arab League.
- Strong presence in Lebanon.
- Possible moves toward normalized relations with Israel.
- History of support of terrorist groups.

Nonmilitary:

- Non-Aligned Movement.

Tunisia

UN voting coincidence and freedom rating

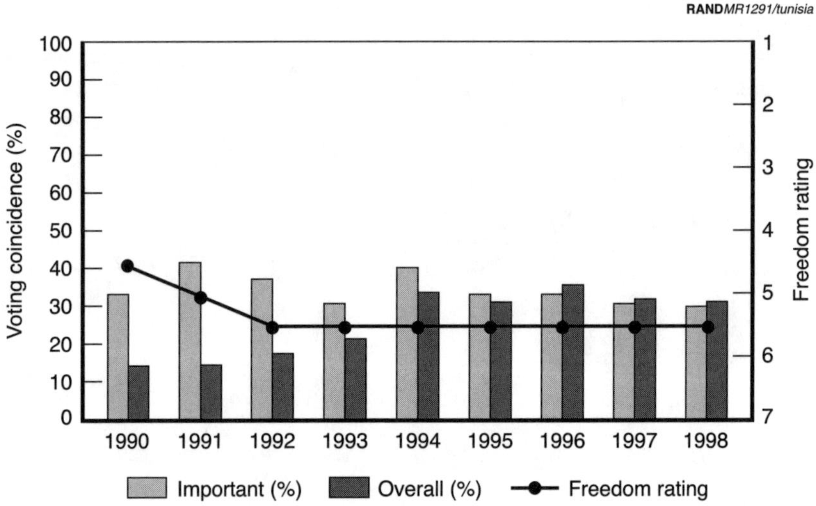

Participation in recent major operations

Desert Storm	Restore Hope	Uphold Democracy	Deliberate Force	Joint Endeavor	Allied Force	Joint Guardian
	✓					

Force size and defense spending

Ground force size	Total force size	Defense spending per soldier
27,000	35,000	approximately $10,000–$11,000

Alignment

Military:

- Arab League.
- Improving relations with Algeria.
- Officer training in France, Egypt, Morocco, U.S.
- African Crisis Response Initiative.

Nonmilitary:

- Non-Aligned Movement.
- OAU.
- Arab Maghreb Union.

Defense and military agreements with the United States

- 1974, military assistance (2 agreements).
- 1985, cooperation in military medicine.
- 1992, officer exchange.
- 1994, logistic support.

Defense ties with the United States

Some officer training in U.S. African Crisis Response Initiative.

Turkey

UN voting coincidence and freedom rating

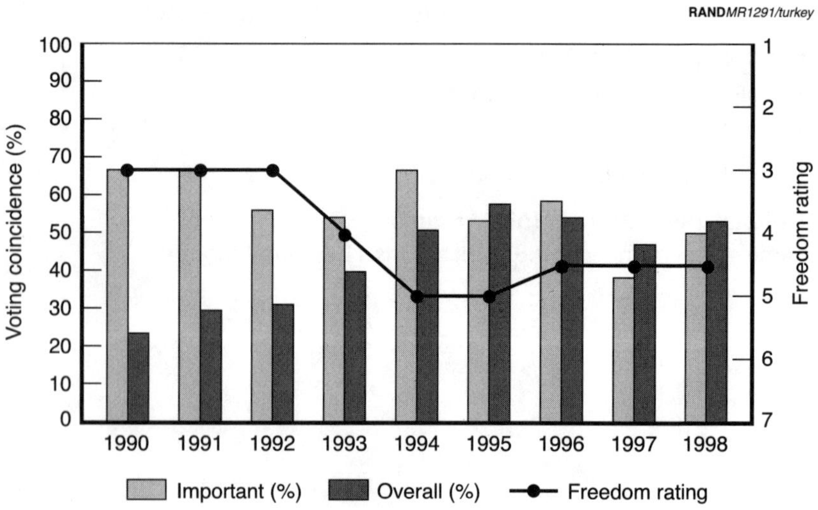

Participation in recent major operations

Desert Storm	Restore Hope	Uphold Democracy	Deliberate Force	Joint Endeavor	Allied Force	Joint Guardian
	✓		✓	✓	✓	✓

Force size and defense spending

Ground force size	Total force size	Defense spending per soldier
525,000	639,000	approximately $12,000–$13,000

Alignment

Military:

- NATO.
- WEU associate member.
- Southeast European Brigade with Albania, Bulgaria, Romania, Greece, Macedonia, Italy.
- Increasing ties with Israel, Jordan.

- Georgian and Azeri forces will serve in Kosovo under Turkish command.

Nonmilitary:

- EU associate.

Defense and military agreements with the United States

- 1954, implementation of NATO SOFA.
- Range of cooperation, weapons production, military assistance, etc. Agreements and arrangements through the present.
- March 1980, defense and economic cooperation agreement established a new framework for U.S. military activities in Turkey and committed the United States to "best efforts" in providing defense support to the Turkish armed forces.

Defense ties with the United States

NATO member. U.S./NATO presence in country. Good ties with U.S. in alliance and bilateral frameworks.

United Arab Emirates

UN voting coincidence and freedom rating

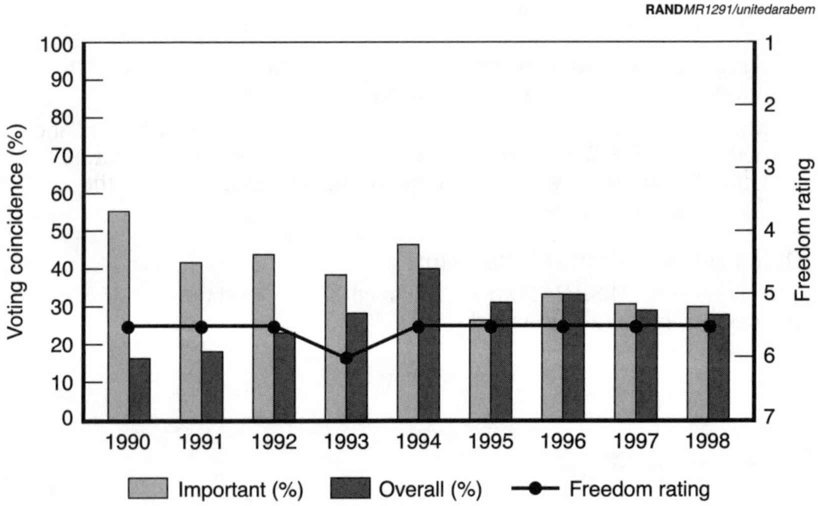

Participation in recent major operations

Desert Storm	Restore Hope	Uphold Democracy	Deliberate Force	Joint Endeavor	Allied Force	Joint Guardian
✓	✓					✓

Force size and defense spending

Ground force size	Total force size	Defense spending per soldier
59,000	64,500	approximately $38,000

Alignment

Military:

- Gulf Cooperation Council.
- Arab League.
- Has sought to rely on the GCC, Syria, Egypt, U.S., other Western allies for security.
- Military ties with Egypt include training in Egypt.

- Training in France and U.K.
- Morocco has loaned out training teams.
- Considerable foreign expatriate participation in military (Egyptian, Moroccan, Baluchi, Sudanese, U.K.).
- French and U.K. forces train on territory.
- Russian naval visits.
- 1997 security agreement with U.K.

Nonmilitary:

- Non-Aligned Movement.

Defense and military agreements with the United States

- 1975, weapons procurement.
- 1987, security of military information.

Defense ties with the United States

U.S. forces train on territory.

Yemen

UN voting coincidence and freedom rating

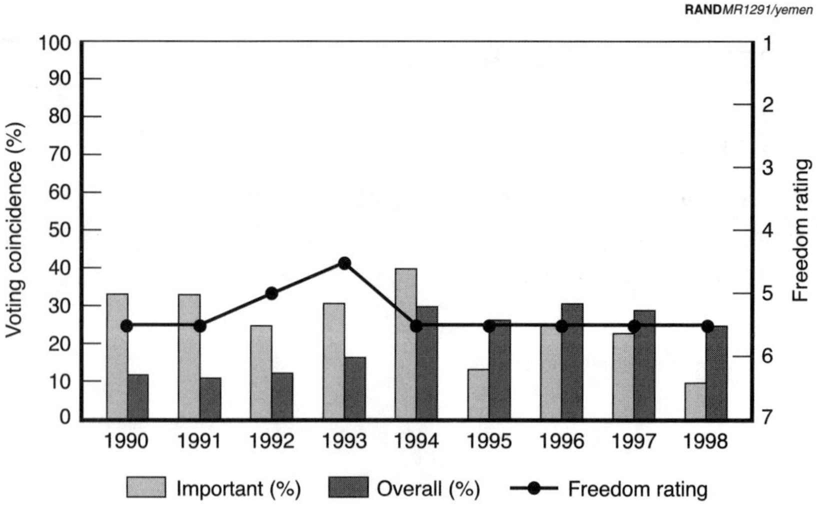

Participation in recent major operations

Desert Storm	Restore Hope	Uphold Democracy	Deliberate Force	Joint Endeavor	Allied Force	Joint Guardian

Force size and defense spending

Ground force size	Total force size	Defense spending per soldier
61,000	66,300	approximately $6,000

Alignment

Military:

- Arab League.
- Continued ties with Iraq.
- Reestablished diplomatic relations with Kuwait.
- U.K. may provide mine-clearing assistance.

Nonmilitary:

- Non-Aligned Movement.

Defense and military agreements with the United States

- 1987, IMET.

Defense ties with the United States

Improving. IMET recently resumed.

Africa (25 countries)

	Important Votes	Overall Votes	Freedom Rating	Number of Ops	Ground Strength	Per-Soldier Spending	Ties & Alignment
Botswana	0.49	0.30	1.94	1	7,000	$21,000–$32,000	High
Burkina Faso	0.29	0.25	4.78	0	5,600	$12,000	Medium
Cameroon	0.31	0.25	5.83	0	11,500	$18,000	Medium
Chad	0.30	0.18	5.67	0	25,000	$2,000	Medium
Cote d'Ivoire	0.30	0.26	5.22	0	6,800	$12,000	Medium
Djibouti	0.38	0.26	5.67	1	8,000	$2,000	Low
Eritrea	0.18	0.24	5.17	0	>20,000	$1,000	Low
Ethiopia	0.37	0.27	5.11	0	100,000	$1,000	Low
Ghana	0.24	0.21	4.33	0	5,000	$7,000–$19,000	Low
Guinea	0.36	0.26	5.50	0	8,500	$5,000	Low
Kenya	0.27	0.26	5.83	0	20,500	$10,000	Medium
Madagascar	0.30	0.22	3.33	0	20,000	$1,000–$2,000	Low
Malawi	0.44	0.26	4.17	0	5,000	. $5,000	Low
Mali	0.34	0.24	3.22	0	7,400	$6,000	Low
Mauritania	0.30	0.24	6.21	0	15,000	$2,000	Low
Mozambique	0.32	0.23	4.39	0	5,000	$10,000–$12,000	Low
Namibia	0.31	0.24	2.50	0	5,700	$15,000	Low
Niger	0.35	0.28	5.00	1	5,200	$4,000	Medium
Nigeria	0.29	0.25	5.72	1	62,000	$26,000	Low
Senegal	0.42	0.28	4.00	1	12,000	$5,000	Low
South Africa	0.26	0.22	2.94	0	54,300	$29,000–$32,000	Low
Tanzania	0.24	0.21	5.28	0	30,000	$4,000	Low
Togo	0.35	0.25	5.61	0	6,500	$4,000	Medium
Uganda	0.25	0.21	4.89	0	47,000	$3,000	Low
Zambia	0.39	0.28	3.83	0	20,000	$3,000	Low
Zimbabwe	0.24	0.23	4.89	1	35,000	$8,000	Low

NOTE: The first three columns report average values, 1990–1998.

KEY	Imp. Votes[a]	Overall Votes[a]	Freedom Rating	Number of Ops.	Ground Strength[b]	Per-Soldier Spending[c]	Alignment
High	0 to 33%	0 to 25%	1 to 3	4 to 7	> 60,000	>$94,000	Frequent regularized contacts, training, agreements, with U.S.
Medium	33 to 60%	25 to 45%	3.1 to 5.5	1 to 3	20,000–60,000	$19,000–$94,000	Infrequent or irregular contacts **or** strong ties with U.S. allies
Low	60 to 100%	45 to 100%	5.5 to 7	0	5,000–19,999	<$19,000	Few, sporadic, or no contacts

[a]For UN votes, the high, medium, and low range cutoffs represent the 0–25th, 25–75th, and 75–100th percentiles, respectively.

[b]Ground-strength low, medium, and high cutoffs are the assessed ability to contribute to multinational operations at a strength of approximately one brigade, one division, or more, respectively. Based on 1997 data.

[c]Defense-spending high, medium, and low cutoffs are the percentage of U.S. per-soldier spending at 20% or less, 20–50%, and over 50%, respectively. Based on 1997 data.

Botswana

UN voting coincidence and freedom rating

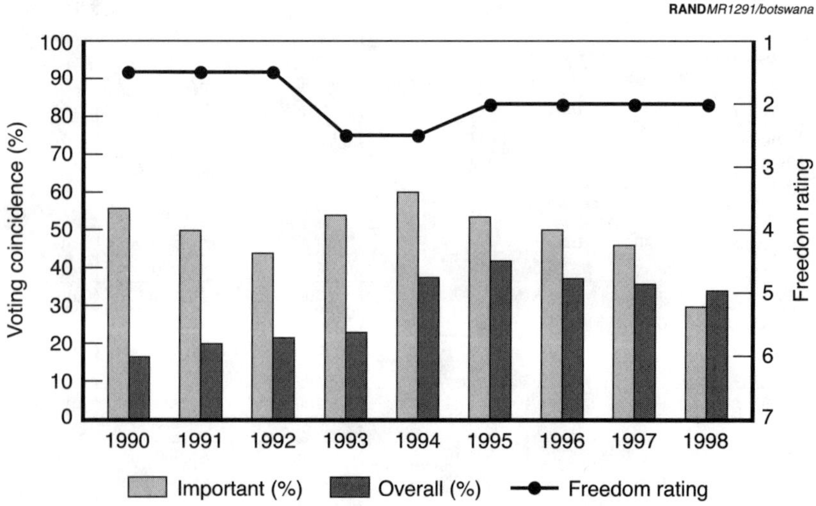

Participation in recent major operations

Desert Storm	Restore Hope	Uphold Democracy	Deliberate Force	Joint Endeavor	Allied Force	Joint Guardian
	✓					

Force size and defense spending

Ground force size	Total force size	Defense spending per soldier
7,000	7,500	approximately $21,000–$32,000

Alignment

Military:

- British have conducted exercises.

Nonmilitary:

- Non-Aligned Movement.
- OAU.
- Commonwealth.

- U.S. training, exercises.

Defense and military agreements with the United States

- 1980, IMET.
- 1992, military assistance (Foreign Assistance Act).

Defense ties with the United States

85% of Botswanan officers trained under IMET as of 1997: about 30/year. Botswana hosts U.S.-run Regional Defense Resource Management seminar with 9 other African countries participating. Air base being used as U.S. listening post. U.S. conducts exercises with Botswanan forces.

Burkina Faso

UN voting coincidence and freedom rating

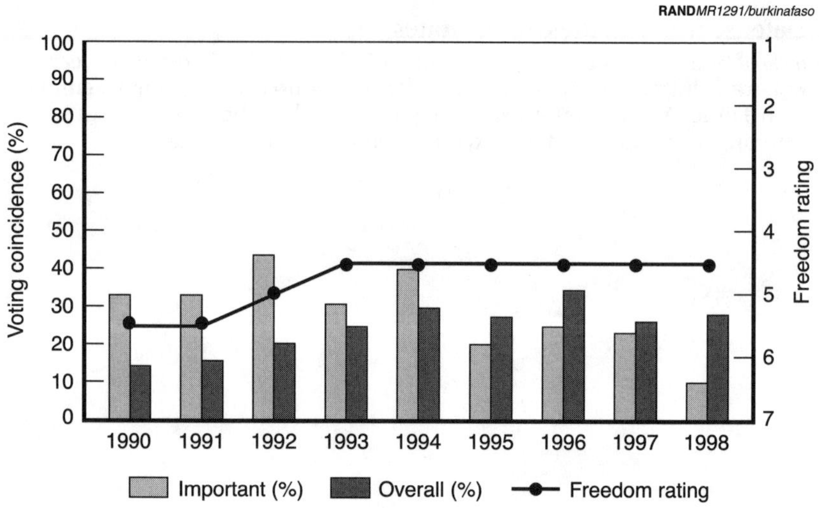

Participation in recent major operations

Desert Storm	Restore Hope	Uphold Democracy	Deliberate Force	Joint Endeavor	Allied Force	Joint Guardian

Force size and defense spending

Ground force size	Total force size	Defense spending per soldier
5,600	5,800	approximately $12,000

Alignment

Military:

- Close ties with France (former colony).
- March 1997, humanitarian exercise conducted with France, Togo, Benin (set up security zone, provide humanitarian aid in an African country with internal strife).
- Has offered France base to train or stage French-sponsored African peacekeeping force.

Nonmilitary:

- Non-Aligned Movement.
- OAU.

Defense and military agreements with the United States

- 1986, IMET.

Defense ties with the United States

Close ties with France.

Cameroon

UN voting coincidence and freedom rating

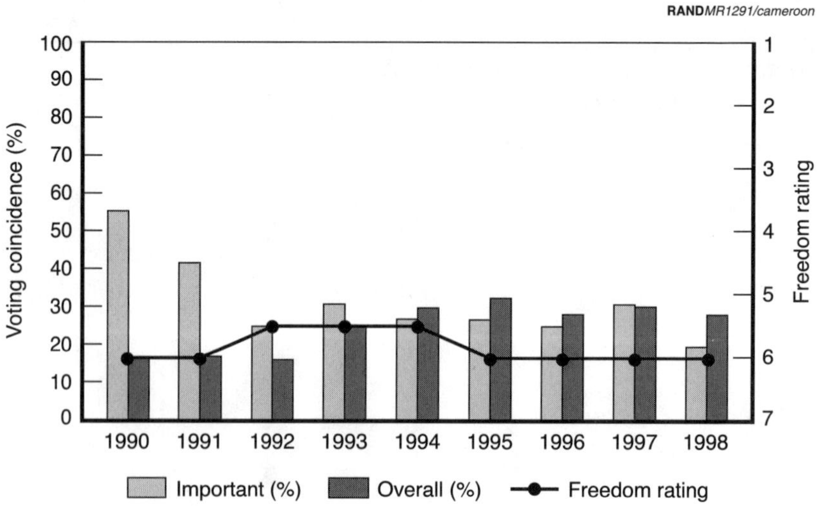

Participation in recent major operations

Desert Storm	Restore Hope	Uphold Democracy	Deliberate Force	Joint Endeavor	Allied Force	Joint Guardian

Force size and defense spending

Ground force size	Total force size	Defense spending per soldier
11,500	13,100	approximately $18,000

Alignment

Military:

- Close links with France: training exercises, etc.

Nonmilitary:

- Non-Aligned Movement; OAU; Commonwealth.

Defense and military agreements with the United States

- 1980, IMET.

Defense ties with the United States

Close ties with France.

Chad

UN voting coincidence and freedom rating

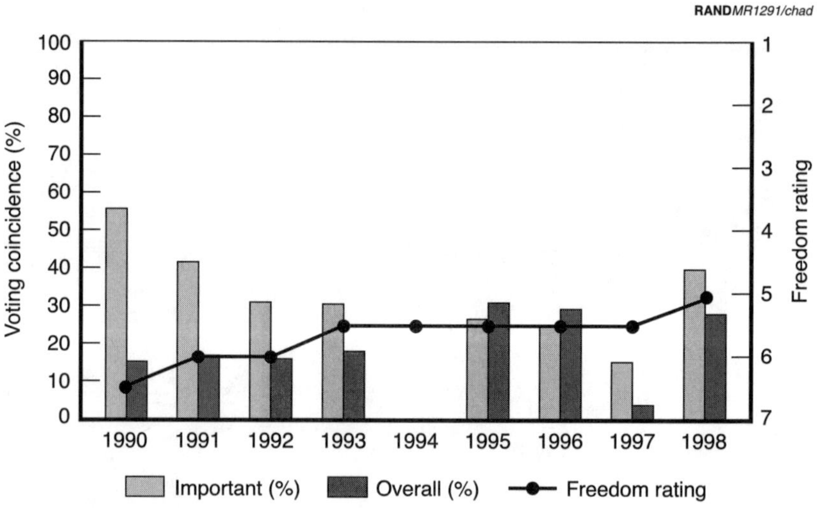

Participation in recent major operations

Desert Storm	Restore Hope	Uphold Democracy	Deliberate Force	Joint Endeavor	Allied Force	Joint Guardian

Force size and defense spending

Ground force size	Total force size	Defense spending per soldier
25,000	25,400	approximately $2,000

Alignment

Military:

- Close relations with France. French forces in country.

Nonmilitary:

- Non-Aligned Movement; OAU.

Defense and military agreements with the United States

- 1983, IMET.
- 1983, mutual defense assistance.
- 1987, SOFA.

Defense ties with the United States

Close ties with France.

Cote d'Ivoire

UN voting coincidence and freedom rating

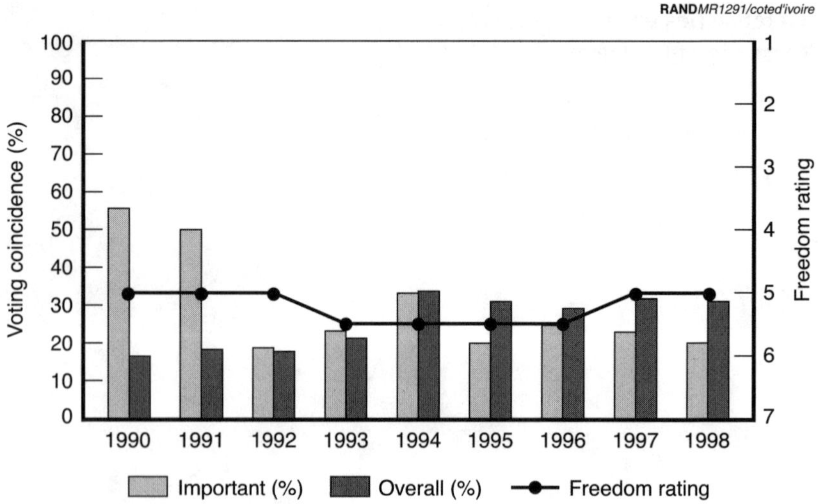

Participation in recent major operations

Desert Storm	Restore Hope	Uphold Democracy	Deliberate Force	Joint Endeavor	Allied Force	Joint Guardian

Force size and defense spending

Ground force size	Total force size	Defense spending per soldier
6,800	8,400	approximately $12,000

Alignment

Military:

- Cote D'Ivoire's French-directed training establishments also train forces from Central African Republic, Gabon, Niger, and Senegal.
- 1997, joint exercise with France to improve capability to participate in multinational peacekeeping.
- French military presence.

- Supports idea of French-led rapid-reaction force, has proposed use of former French bases on territory as training centers or staging posts.
- African Crisis Response Initiative designee.

Nonmilitary:

- Non-Aligned Movement.
- OAU.

Defense and military agreements with the United States

- 1983, IMET.

Defense ties with the United States

African Crisis Response Initiative designee.

Djibouti

UN voting coincidence and freedom rating

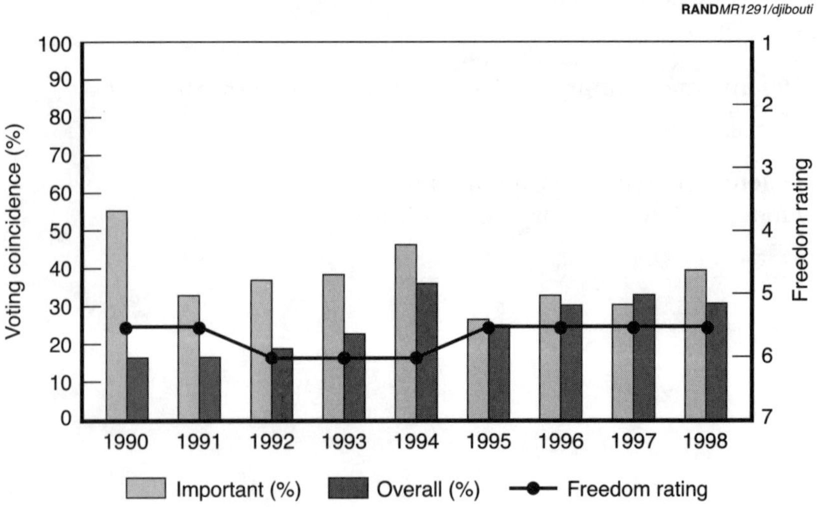

Participation in recent major operations

Desert Storm	Restore Hope	Uphold Democracy	Deliberate Force	Joint Endeavor	Allied Force	Joint Guardian
		✓				

Force size and defense spending

Ground force size	Total force size	Defense spending per soldier
8,000	9,600	approximately $2,000

Alignment

Military:

- Arab League.
- About 150 officers/NCOs per year are trained in France.
- French presence in country.
- 1997, exercise with South Africa, France, Mauritius, Madagascar, Mayotte.

Nonmilitary:

- Non-Aligned Movement; OAU.

Defense and military agreements with the United States

- 1984, IMET.
- 1992, military assistance.

Defense ties with the United States

Ties with France.

Eritrea

UN voting coincidence and freedom rating

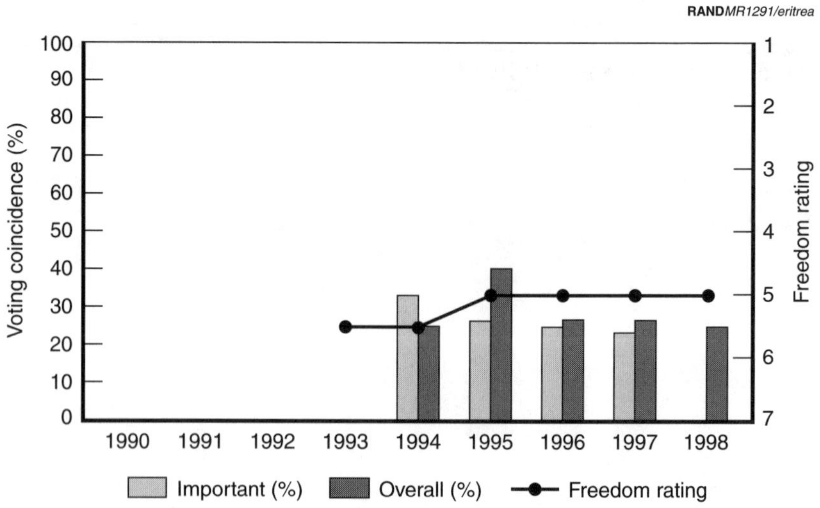

Participation in recent major operations

Desert Storm	Restore Hope	Uphold Democracy	Deliberate Force	Joint Endeavor	Allied Force	Joint Guardian

Force size and defense spending

Ground force size	Total force size	Defense spending per soldier
46,000 (1998)	46,000	approximately $1,000

Alignment

Military:

- Good relations with the United States (had mutual defense treaty 1953–1977), Italy, United Kingdom, Germany, and Norway, all based on aid.
- Close ties with Yemen.

Nonmilitary:

- Non-Aligned Movement.
- OAU.

Defense and military agreements with the United States

- 1993, IMET.
- 1995, military assistance (Foreign Assistance Act).

Ethiopia

UN voting coincidence and freedom rating

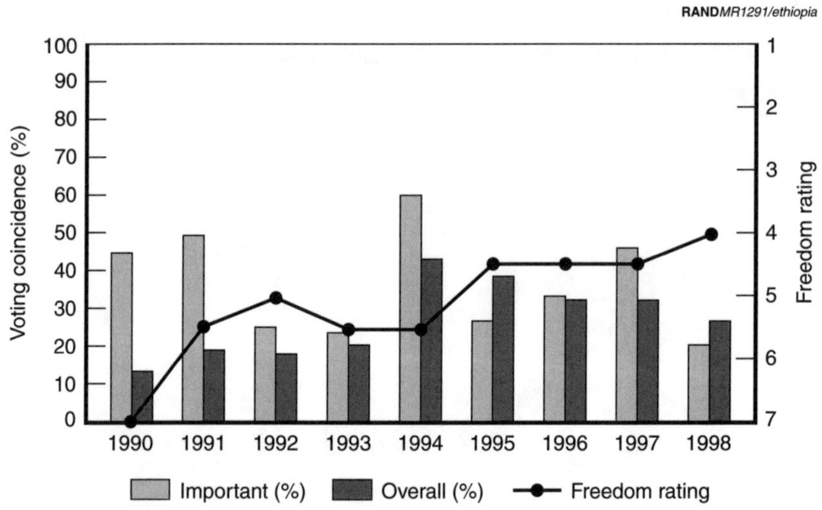

Participation in recent major operations

Desert Storm	Restore Hope	Uphold Democracy	Deliberate Force	Joint Endeavor	Allied Force	Joint Guardian

Force size and defense spending

Ground force size	Total force size	Defense spending per soldier
100,000	120,000	approximately $1,000

Alignment

Military:

- African Crisis Response Initiative.

Nonmilitary:

- Non-Aligned Movement.
- OAU.

Defense and military agreements with the United States

- 1994, SOFA.
- 1997, SOFA.

Ghana

UN voting coincidence and freedom rating

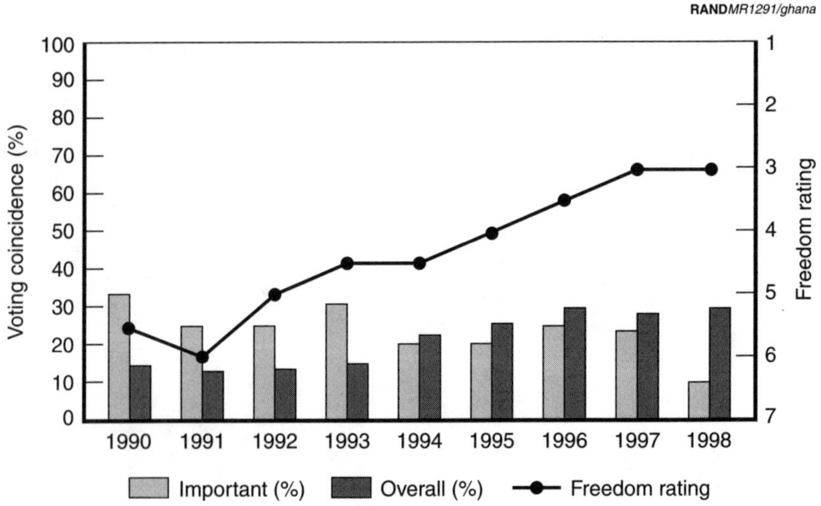

Participation in recent major operations

Desert Storm	Restore Hope	Uphold Democracy	Deliberate Force	Joint Endeavor	Allied Force	Joint Guardian

Force size and defense spending

Ground force size	Total force size	Defense spending per soldier
5,000	7,000	approximately $7,000–$19,000

Alignment

Military:

- U.S. or U.K. special forces may be training Ghanian troops.

Nonmilitary:

- Non-Aligned Movement.
- OAU.
- Commonwealth.

Defense and military agreements with the United States

- 1986, IMET.
- 1998, SOFA (African Crisis Response Initiative).

Defense ties with the United States

African Crisis Response Initiative, U.S. medical personnel provided training and humanitarian relief in 1994.

Guinea

UN voting coincidence and freedom rating

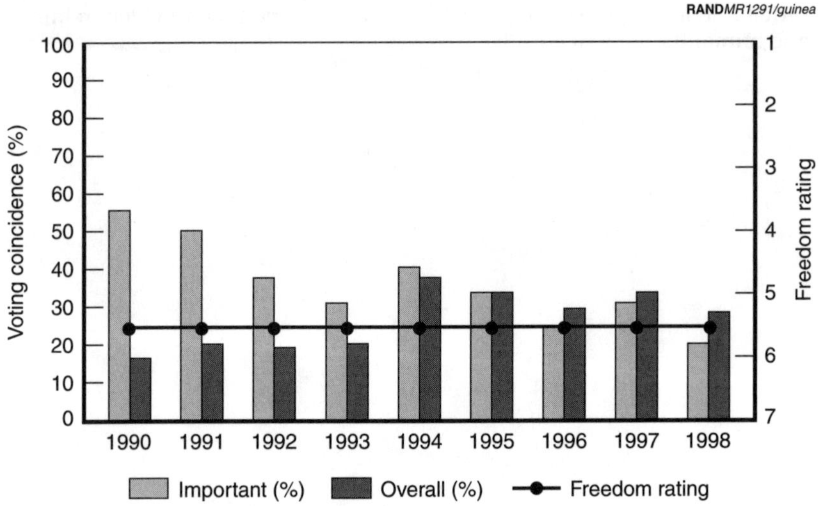

Participation in recent major operations

Desert Storm	Restore Hope	Uphold Democracy	Deliberate Force	Joint Endeavor	Allied Force	Joint Guardian

Force size and defense spending

Ground force size	Total force size	Defense spending per soldier
8,500	9,700	approximately $5,000

Alignment

Nonmilitary:

- Non-Aligned Movement.
- OAU.

Defense and military agreements with the United States

- 1965, military assistance.
- 1983, IMET.

Kenya

UN voting coincidence and freedom rating

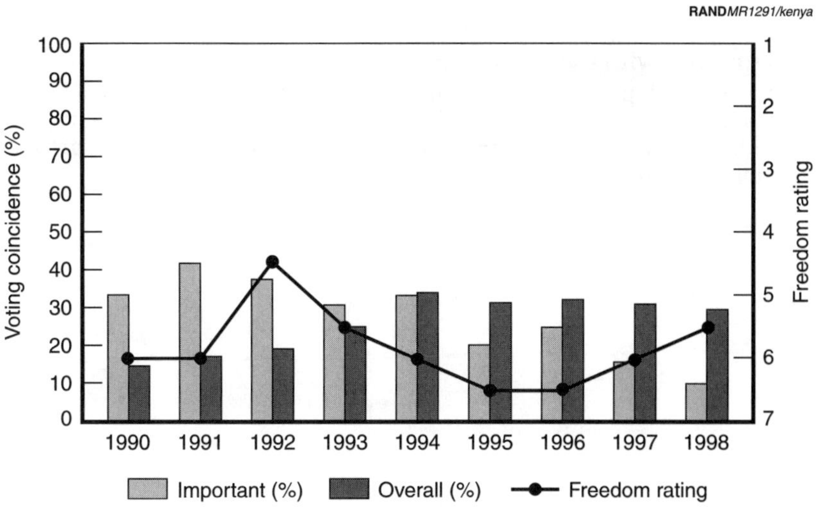

Participation in recent major operations

Desert Storm	Restore Hope	Uphold Democracy	Deliberate Force	Joint Endeavor	Allied Force	Joint Guardian

Force size and defense spending

Ground force size	Total force size	Defense spending per soldier
20,500	24,200	approximately $10,000

Alignment

Military:

- Improving relations with Uganda and Tanzania (together in East African Cooperation group).
- Good relations with Sudan and Ethiopia.

Nonmilitary:

- Non-Aligned Movement.

- OAU.
- Commonwealth.

Defense and military agreements with the United States

- 1976, military assistance and training (International Security Assistance and Arms Export Control Act)

Defense ties with the United States

Long relationship, recently somewhat cooled in aftermath of Cold War. Kenya used as staging point for Gulf War; friction in early efforts in Somalia over U.S. military control of operation. U.S. wants to involve in African Crisis Response Initiative.

Madagascar

UN voting coincidence and freedom rating

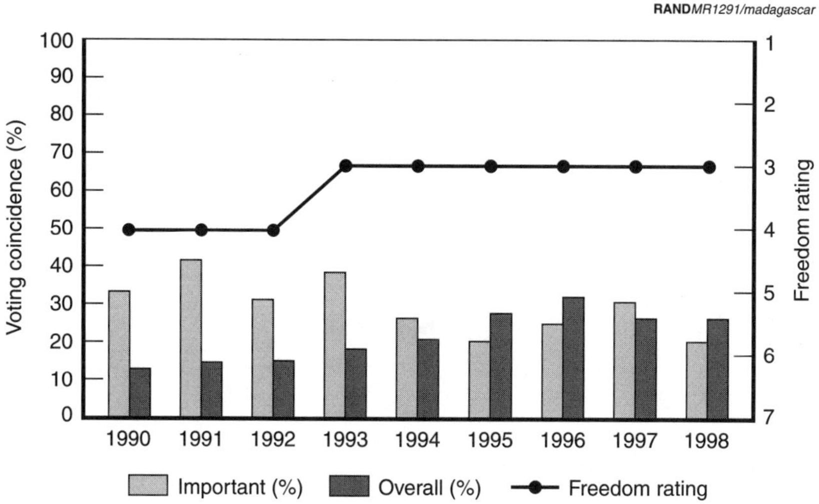

Participation in recent major operations

Desert Storm	Restore Hope	Uphold Democracy	Deliberate Force	Joint Endeavor	Allied Force	Joint Guardian

Force size and defense spending

Ground force size	Total force size	Defense spending per soldier
20,000	21,000	approximately $1,000–$2,000

Alignment

Military:

- Active relationships with France, Germany, Switzerland, Russia, Japan, India, and China.
- 1997, exercise with France, Mauritius, Djibouti, Mayotte, South Africa.
- Ties with South Africa, South Africa carries out some training.

Nonmilitary:

- Non-Aligned Movement; OAU.

Defense and military agreements with the United States

- 1984, IMET.

Defense ties with the United States

Ties with allies.

Malawi

UN voting coincidence and freedom rating

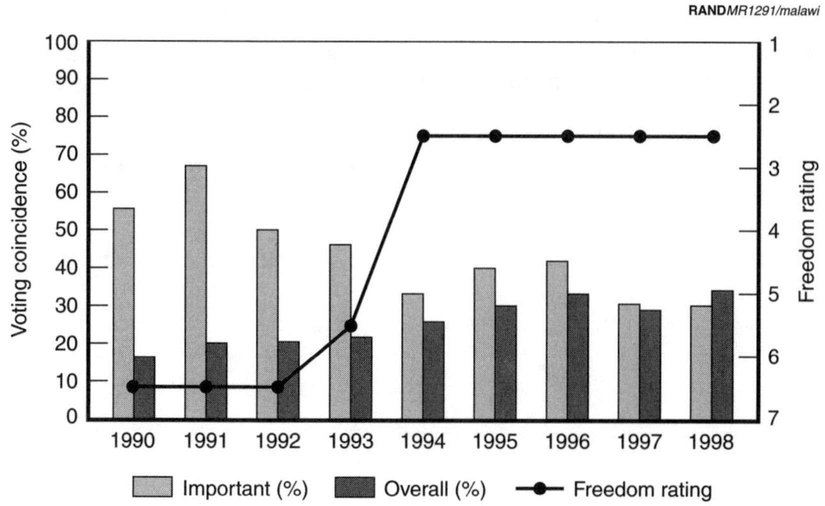

Participation in recent major operations

Desert Storm	Restore Hope	Uphold Democracy	Deliberate Force	Joint Endeavor	Allied Force	Joint Guardian

Force size and defense spending

Ground force size	Total force size	Defense spending per soldier
5,000	5,000	approximately $5,000

Alignment

Military:

- Training in South Africa, U.K.
- Assistance from Taiwan.
- African Crisis Response Initiative.

Nonmilitary:

- Non-Aligned Movement.

- OAU.
- Commonwealth.

Defense and military agreements with the United States

- 1980, IMET.
- 1992, military assistance (Foreign Assistance Act).
- 1997, military assistance (commodities, services, and associated military education and training for African Crisis Response Initiative).

Defense ties with the United States

African Crisis Response Initiative, U.S. to provide training related to the initiative.

Mali

UN voting coincidence and freedom rating

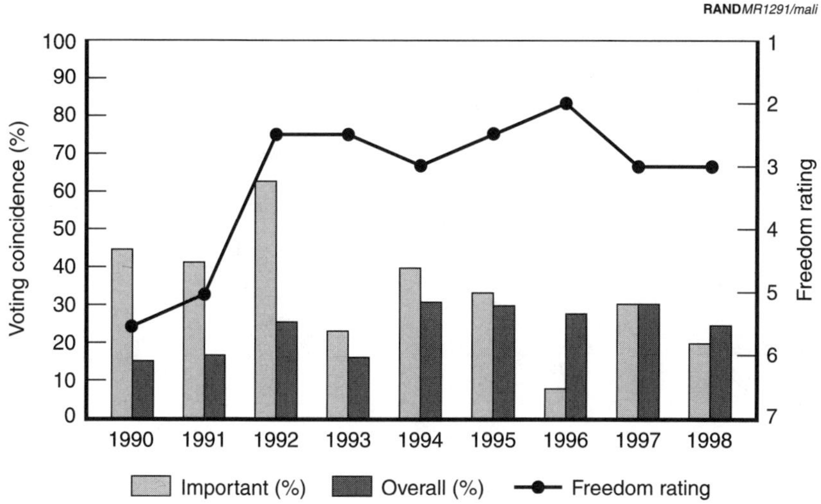

Participation in recent major operations

Desert Storm	Restore Hope	Uphold Democracy	Deliberate Force	Joint Endeavor	Allied Force	Joint Guardian

Force size and defense spending

Ground force size	Total force size	Defense spending per soldier
7,400	7,400	approximately $6,000

Alignment

Nonmilitary:

- Non-Aligned Movement.
- OAU.
- African Crisis Response Initiative.

Defense and military agreements with the United States

- 1961, military assistance (related 1972).

- 1967, C-47 delivery.
- 1984, IMET.

Defense ties with the United States
African Crisis Response Initiative.

Mauritania

UN voting coincidence and freedom rating

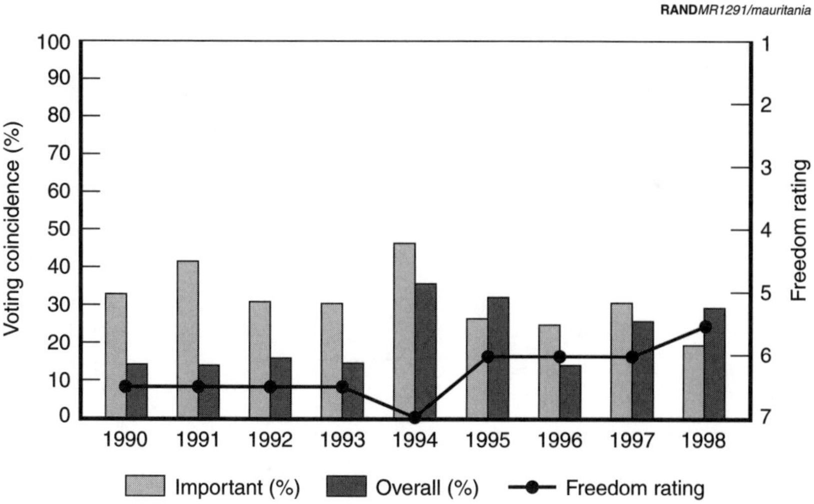

Participation in recent major operations

Desert Storm	Restore Hope	Uphold Democracy	Deliberate Force	Joint Endeavor	Allied Force	Joint Guardian

Force size and defense spending

Ground force size	Total force size	Defense spending per soldier
15,000	15,700	approximately $2,000

Alignment

Military:

- Arab League.
- French training.

Nonmilitary:

- Non-Aligned Movement.
- OAU.

- Arab Maghreb Union.

Defense and military agreements with the United States

- 1984, IMET.

Mozambique

UN voting coincidence and freedom rating

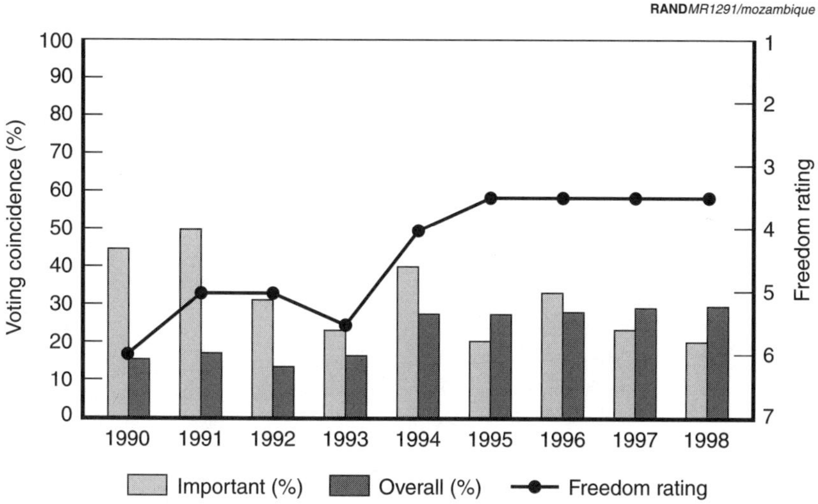

Participation in recent major operations

Desert Storm	Restore Hope	Uphold Democracy	Deliberate Force	Joint Endeavor	Allied Force	Joint Guardian

Force size and defense spending

Ground force size	Total force size	Defense spending per soldier
5,000	6,100	approximately $10,000–$12,000

Alignment

Military:

- 1996, defense cooperation agreement with South Africa. Will cover training, field exercises, intelligence sharing, and military medical cooperation.
- U.K. training.

Nonmilitary:

- Non-Aligned Movement.

- Commonwealth.
- OAU.

Defense and military agreements with the United States

- 1985, IMET.
- 1994, military assistance (Foreign Assistance Act).

Namibia

UN voting coincidence and freedom rating

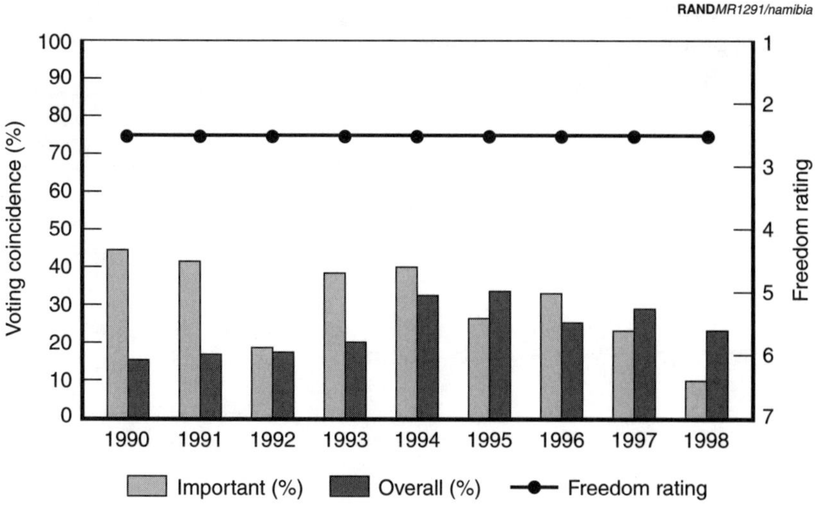

Participation in recent major operations

Desert Storm	Restore Hope	Uphold Democracy	Deliberate Force	Joint Endeavor	Allied Force	Joint Guardian

Force size and defense spending

Ground force size	Total force size	Defense spending per soldier
5,700	5,800	approximately $15,000

Alignment

Military:

- Warm relations with Zambia.
- Close ties with South Africa.
- Developing ties with Germany and Japan.
- South Africa has assisted in training.
- 1997, agreement with Russia covering training and support for Russian equipment.

Nonmilitary:

- Non-Aligned Movement.
- OAU.
- Commonwealth.

Defense ties with the United States

U.S. to provide mine-clearing training.

Niger

UN voting coincidence and freedom rating

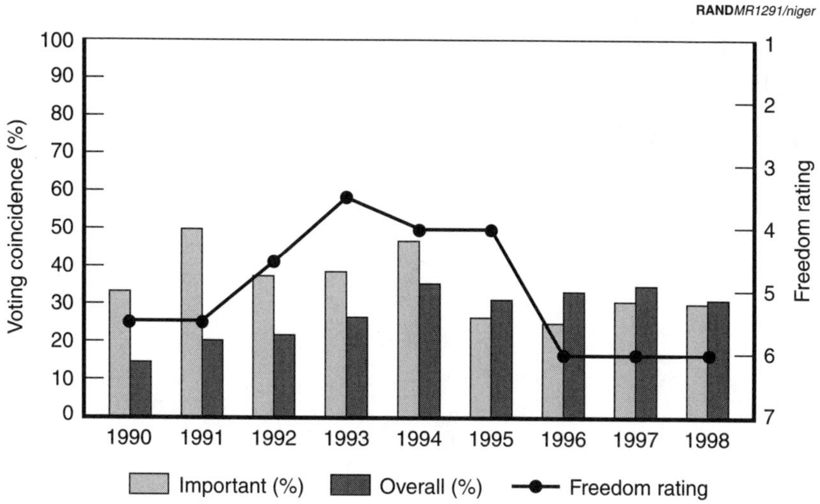

Participation in recent major operations

Desert Storm	Restore Hope	Uphold Democracy	Deliberate Force	Joint Endeavor	Allied Force	Joint Guardian
✓						

Force size and defense spending

Ground force size	Total force size	Defense spending per soldier
5,200	5,300	approximately $4,000

Alignment

Military:

- Special relationship with France (former colony).
- Cote D'Ivoire's French-directed training establishments also train forces from Niger.
- Security guaranteed by France.
- French training assistance.

Nonmilitary:

- Non-Aligned Movement.
- OAU.

Defense and military agreements with the United States

- 1962, military assistance (equipment, materials, and services; 1992, agreement on grant assistance).
- 1980, IMET.

Defense ties with the United States

Close ties with France.

Nigeria

UN voting coincidence and freedom rating

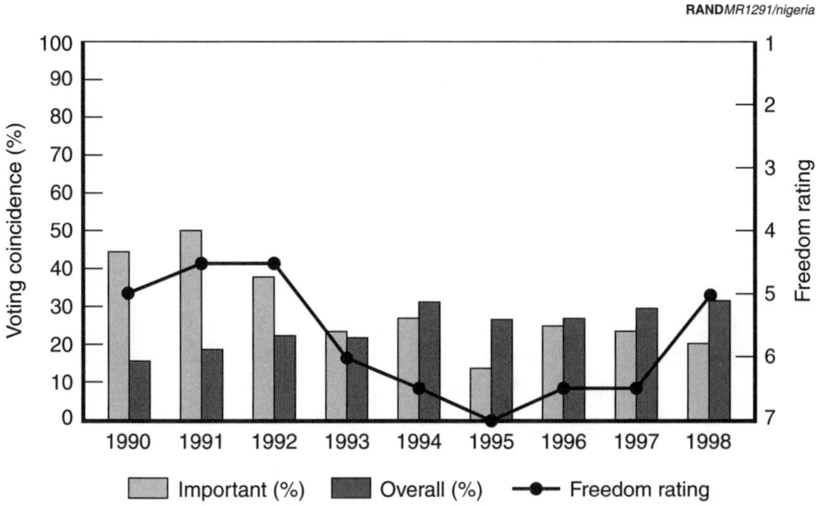

Participation in recent major operations

Desert Storm	Restore Hope	Uphold Democracy	Deliberate Force	Joint Endeavor	Allied Force	Joint Guardian
	✓					

Force size and defense spending

Ground force size	Total force size	Defense spending per soldier
62,000	77,000	approximately $26,000

Alignment

Military:

- Leads ECOMOG.

Nonmilitary:

- Non-Aligned Movement.
- OAU.
- Suspended from the Commonwealth.

Defense and military agreements with the United States

- 1972, military assistance (+ related).
- 1986, IMET.
- 1996, support for ECOMOG.

Senegal

UN voting coincidence and freedom rating

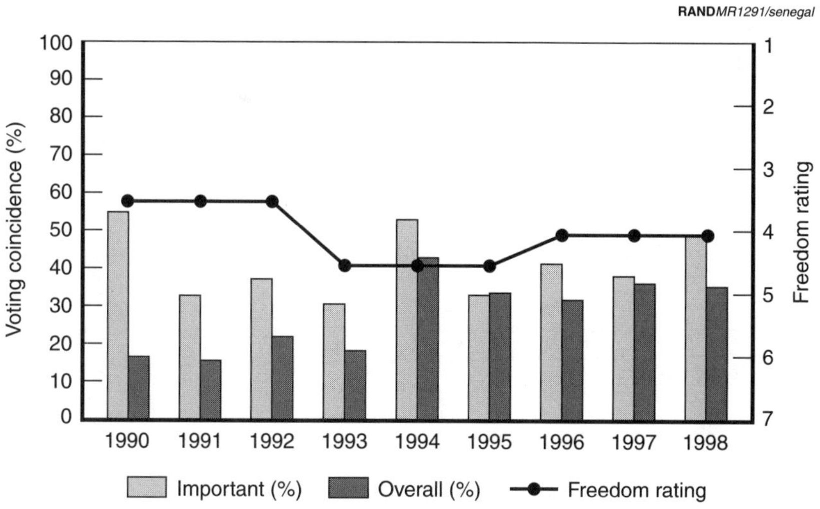

Participation in recent major operations

Desert Storm	Restore Hope	Uphold Democracy	Deliberate Force	Joint Endeavor	Allied Force	Joint Guardian
✓						

Force size and defense spending

Ground force size	Total force size	Defense spending per soldier
12,000	13,400	approximately $5,000

Alignment

Military:

- Close relations with France (former French colony).
- Has offered France use of former French bases on territory for training or staging of proposed French-sponsored African peacekeeping force.
- Cote D'Ivoire's French-directed training establishments also train forces from Senegal.
- French presence.

- African Crisis Response Initiative.

Nonmilitary:

- Non-Aligned Movement.
- OAU.

Defense and military agreements with the United States

- 1962, military assistance (equipment, materials, services).
- 1983, IMET.
- 1997, military assistance (commodities, services, and associated military education and training for African Crisis Response Initiative).

Defense ties with the United States

African Crisis Response Initiative (U.S.) and related training, ties with France.

South Africa

UN voting coincidence and freedom rating

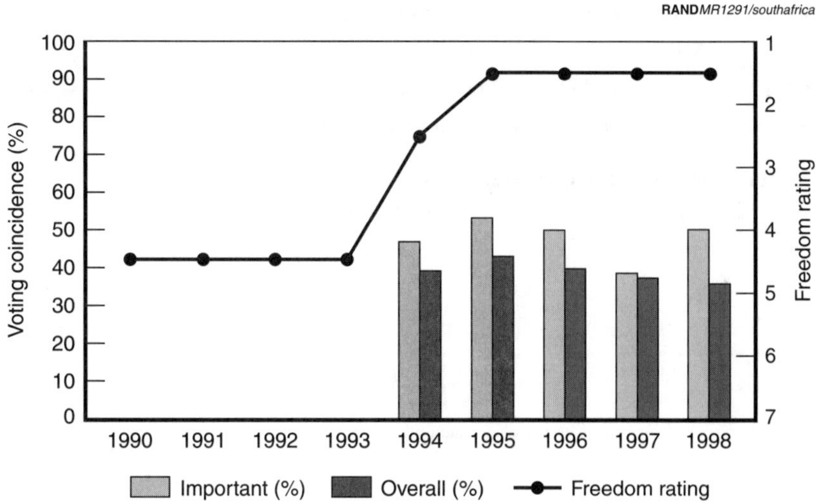

Participation in recent major operations

Desert Storm	Restore Hope	Uphold Democracy	Deliberate Force	Joint Endeavor	Allied Force	Joint Guardian

Force size and defense spending

Ground force size	Total force size	Defense spending per soldier
54,300	79,400	approximately $29,000–$32,000

Alignment

Military:

- 1997, exercise with France, Mauritius, Madagascar, Mayotte, Djibouti.
- U.K. training team assists with military integration.
- MOU with Canada: defense industry cooperation.

Nonmilitary:

- Non-Aligned Movement.

- OAU.
- Commonwealth.

Defense and military agreements with the United States

- 1951, relating to mutual defense assistance.
- 1995, military assistance (Foreign Assistance Act).
- 1994, IMET.
- 1997, defense trade control cooperation.
- 1998, protection of classified military information.

Defense ties with the United States

Sporadic ties. U.S. wants to involve in African Crisis Response Initiative.

Tanzania

UN voting coincidence and freedom rating

RANDMR1291/tanzania

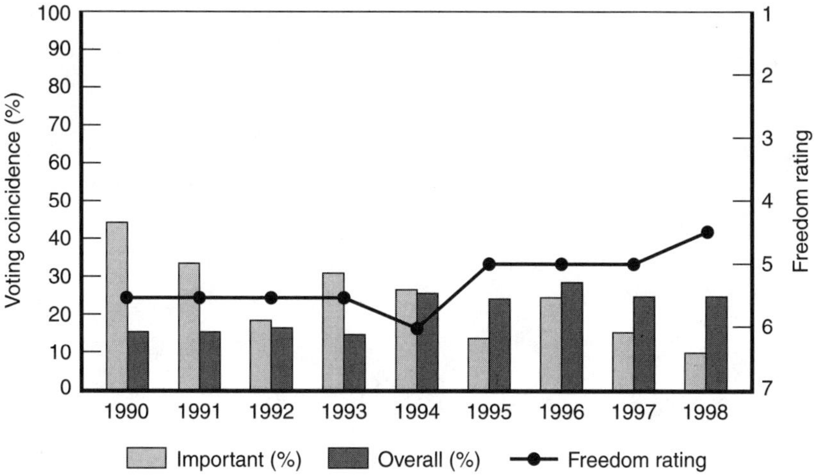

Participation in recent major operations

Desert Storm	Restore Hope	Uphold Democracy	Deliberate Force	Joint Endeavor	Allied Force	Joint Guardian

Force size and defense spending

Ground force size	Total force size	Defense spending per soldier
30,000	34,600	approximately $4,000

Alignment

Military:

- Close ties with neighboring Uganda, Kenya, Zambia, and Mozambique.
- East African cooperation group launched with Kenya and Uganda.
- Some Nigerian and Ghanian training.
- Some ties with South Africa.

Nonmilitary:

- Non-Aligned Movement.
- OAU.
- Commonwealth.

Togo

UN voting coincidence and freedom rating

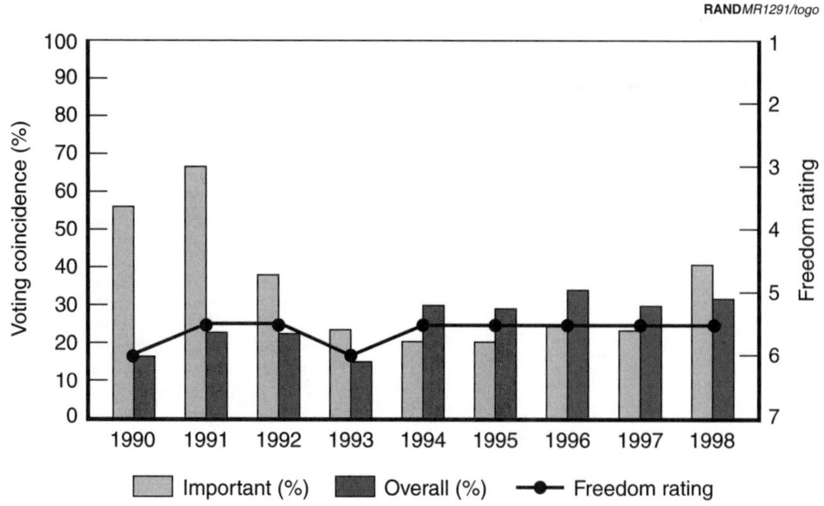

Participation in recent major operations

Desert Storm	Restore Hope	Uphold Democracy	Deliberate Force	Joint Endeavor	Allied Force	Joint Guardian

Force size and defense spending

Ground force size	Total force size	Defense spending per soldier
6,500	7,000	approximately $4,000

Alignment

Military:

- Generally good relations with neighbors.
- Strong historical and cultural ties with Europe, especially France and Germany.
- March 1997, humanitarian exercise conducted with France, Benin, Burkina Faso (set up security zone, provide humanitarian aid in an African country with internal strife).

- Has offered former French bases on territories for training or staging of proposed French-sponsored African peacekeeping force.
- French military assistance: training and exercises.

Nonmilitary:

- Non-Aligned Movement
- OAU

Defense and military agreements with the United States

- 1980, IMET.

Uganda

UN voting coincidence and freedom rating

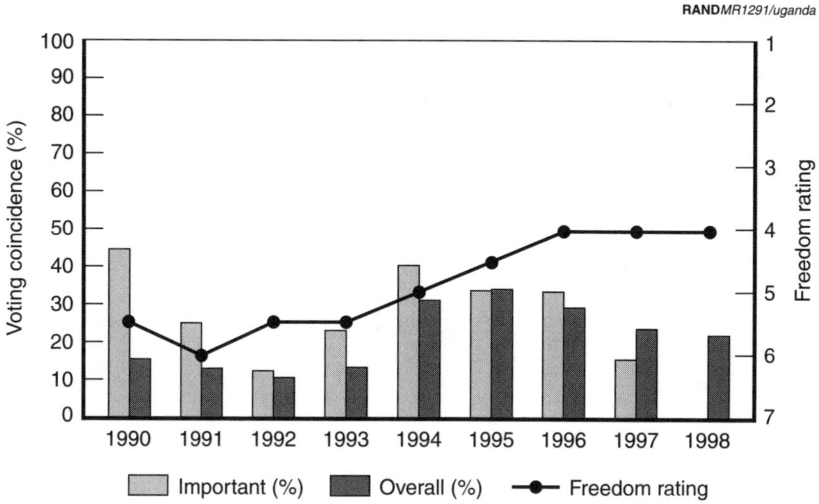

Participation in recent major operations

Desert Storm	Restore Hope	Uphold Democracy	Deliberate Force	Joint Endeavor	Allied Force	Joint Guardian

Force size and defense spending

Ground force size	Total force size	Defense spending per soldier
47,000	55,000	approximately $3,000

Alignment

Military:

- East African Cooperation group with Kenya and Tanzania.
- African Crisis Response Initiative.

Nonmilitary:

- Non-Aligned Movement.
- OAU.

- Commonwealth.

Defense and military agreements with the United States

- 1981, IMET.
- 1994, military assistance (Foreign Assistance Act).
- 1994, SOFA.

Defense ties with the United States

African Crisis Response Initiative with related training.

Zambia

UN voting coincidence and freedom rating

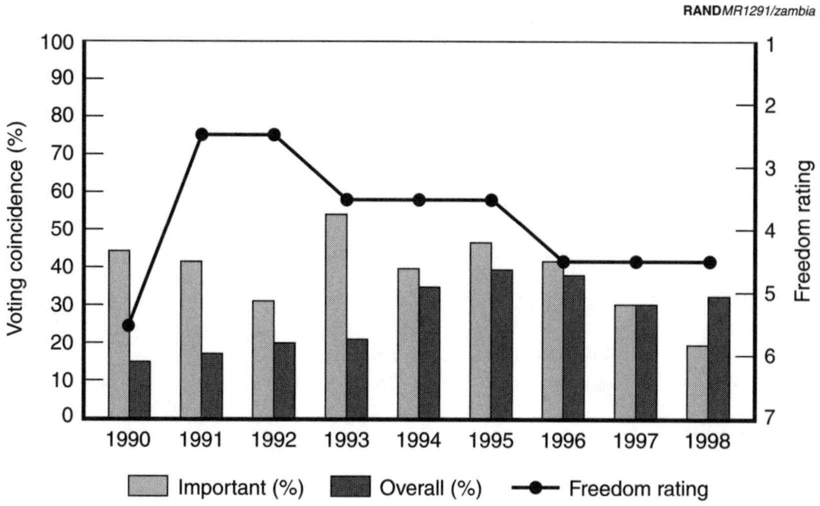

Participation in recent major operations

Desert Storm	Restore Hope	Uphold Democracy	Deliberate Force	Joint Endeavor	Allied Force	Joint Guardian

Force size and defense spending

Ground force size	Total force size	Defense spending per soldier
20,000	21,600	approximately $3,000

Alignment

Military:

- South Africa may provide advisers as ties grow.
- Improved relations with Botswana.
- Improved but still problematical relations with Zimbabwe.

Nonmilitary:

- Non-Aligned Movement.

- OAU.
- Commonwealth.

Defense and military agreements with the United States

- 1992, IMET.

Zimbabwe

UN voting coincidence and freedom rating

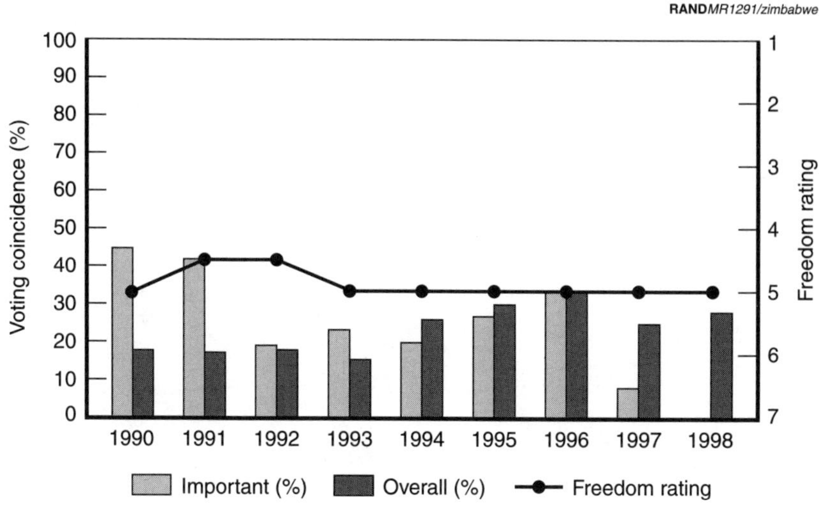

Participation in recent major operations

Desert Storm	Restore Hope	Uphold Democracy	Deliberate Force	Joint Endeavor	Allied Force	Joint Guardian
	✓					

Force size and defense spending

Ground force size	Total force size	Defense spending per soldier
35,000	39,000	approximately $8,000

Alignment

Military:

- Denmark funding peacekeeping center in Zimbabwe.
- German assistance.
- Small British training team in country.

Nonmilitary:

- Non-Aligned Movement; OAU; Commonwealth.

Defense and military agreements with the United States

- 1981, IMET.

Defense ties with the United States

Wants to involve in African Crisis Response Initiative, ties with allies.

North and Central America (7 countries)

	Important Votes	Overall Votes	Freedom Rating	Number of Ops	Ground Strength	Per-Soldier Spending	Ties & Alignment
Canada	0.70	0.51	1.00	6	21,900	$126,000–$130,000	High
Dominican Republic	0.38	0.17	2.89	0	15,000	$5,000	Low
El Salvador	0.59	0.30	3.00	0	25,700	$3,000–$6,000	High
Guatemala	0.49	0.29	4.00	1	38,500	$2,000–$4,000	High
Honduras	0.57	0.32	2.72	1	16,000	$1,000–$5,000	High
Mexico	0.47	0.27	3.78	0	130,000	$7,000–$21,000	Low
Nicaragua	0.55	0.31	3.44	0	15,000	$2,000	Low

NOTE: The first three columns report average values, 1990–1998.

KEY	Imp. Votes[a]	Overall Votes[a]	Freedom Rating	Number of Ops.	Ground Strength[b]	Per-Soldier Spending[c]	Alignment
High	0 to 33%	0 to 25%	1 to 3	4 to 7	>60,000	>$94,000	Frequent regularized contacts, training, agreements, with U.S.
Medium	33 to 60%	25 to 45%	3.1 to 5.5	1 to 3	20,000–60,000	$19,000–$94,000	Infrequent or irregular contacts **or** strong ties with U.S. allies
Low	60 to 100%	45 to 100%	5.5 to 7	0	5,000–19,999	<$19,000	Few, sporadic, or no contacts

[a]For UN votes, the high, medium, and low range cutoffs represent the 0–25th, 25–75th, and 75–100th percentiles, respectively.

[b]Ground-strength low, medium, and high cutoffs are the assessed ability to contribute to multinational operations at a strength of approximately one brigade, one division, or more, respectively. Based on 1997 data.

[c]Defense-spending high, medium, and low cutoffs are the percentage of U.S. per-soldier spending at 20% or less, 20–50%, and over 50%, respectively. Based on 1997 data.

Canada

UN voting coincidence and freedom rating

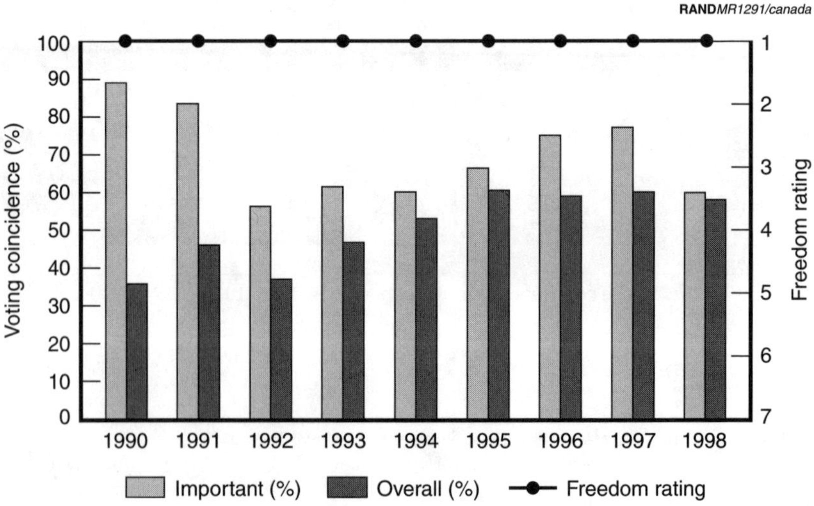

Participation in recent major operations

Desert Storm	Restore Hope	Uphold Democracy	Deliberate Force	Joint Endeavor	Allied Force	Joint Guardian
✓	✓	✓		✓	✓	✓

Force size and defense spending

Ground force size	Total force size	Defense spending per soldier
21,900	61,600	approximately $126,000–$130,000

Alignment

Military:

- NATO.
- MOU to enhance defense industrial cooperation with South Africa.

Nonmilitary:

- OAS.
- Commonwealth.

- ARF (ASEAN dialogue partner).

Defense and military agreements with the United States

- 1940, Permanent Joint Board on Defense, and Military Coordination Committee.
- 1940–present, range of agreements relating to naval and air bases; defense installations; application of NATO SOFA; early warning; etc.

Defense ties with the United States

Canada shares air defense with U.S. (NORAD). Modeled on U.S. and U.K. forces. Close alliance ties with regular exercises, meetings, etc.

Dominican Republic

UN voting coincidence and freedom rating

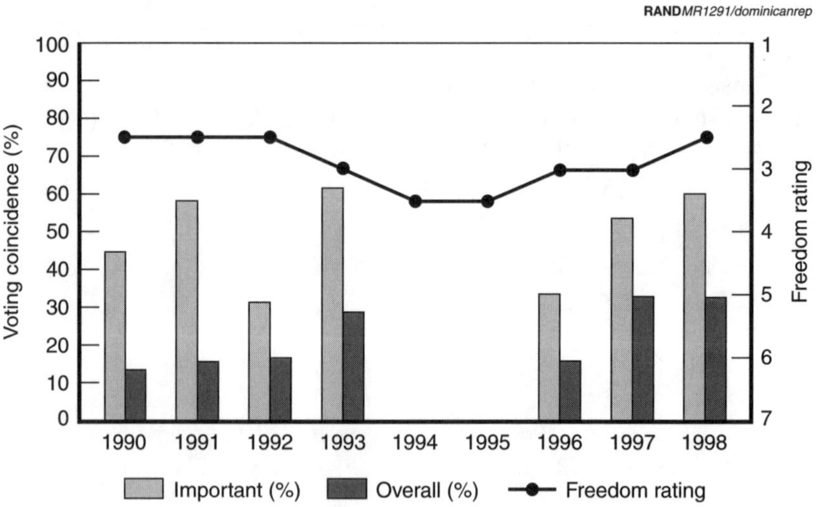

Participation in recent major operations

Desert Storm	Restore Hope	Uphold Democracy	Deliberate Force	Joint Endeavor	Allied Force	Joint Guardian

Force size and defense spending

Ground force size	Total force size	Defense spending per soldier
15,000	24,500	approximately $5,000

Alignment

Military:

- Rio Treaty.

Nonmilitary:

- OAS.

Defense and military agreements with the United States

- 1953, military assistance; terminated 1961 except for certain provisions.
- 1962, military assistance, with related follow-ons in 1972, 1974.
- 1988, SOFA.

Defense ties with the United States

U.S. Command and Staff training under 1962 agreement in the U.S. and Panama Canal zone. Exercises, good cooperation.

El Salvador

UN voting coincidence and freedom rating

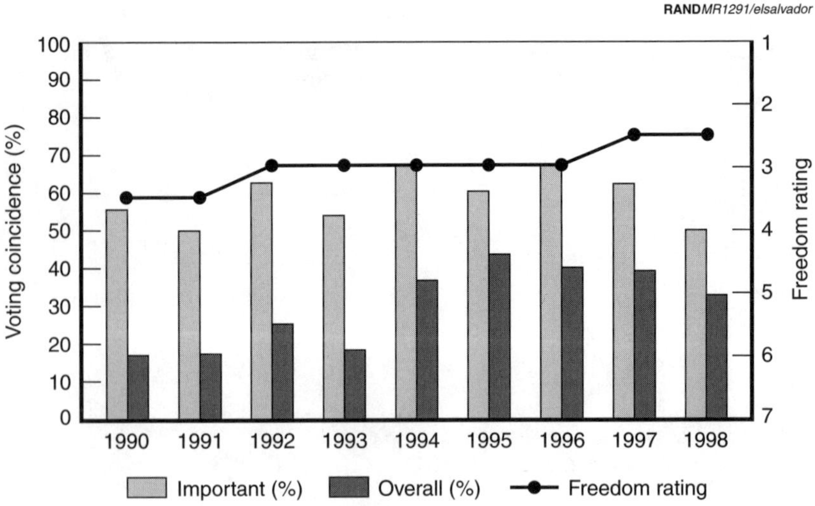

Participation in recent major operations

Desert Storm	Restore Hope	Uphold Democracy	Deliberate Force	Joint Endeavor	Allied Force	Joint Guardian

Force size and defense spending

Ground force size	Total force size	Defense spending per soldier
25,700	28,400	approximately $3,000–$6,000

Alignment

Military:

- Rio Treaty.
- 1997, computer-simulated exercise run by U.S. included soldiers from U.S., Guatemala, Honduras, El Salvador, and Nicaragua.
- Close ties with U.S.

Nonmilitary:

- OAS.

Defense and military agreements with the United States

- 1962, military assistance (defense articles and services).

Defense ties with the United States

Very close. Exercises, etc.

Guatemala

UN voting coincidence and freedom rating

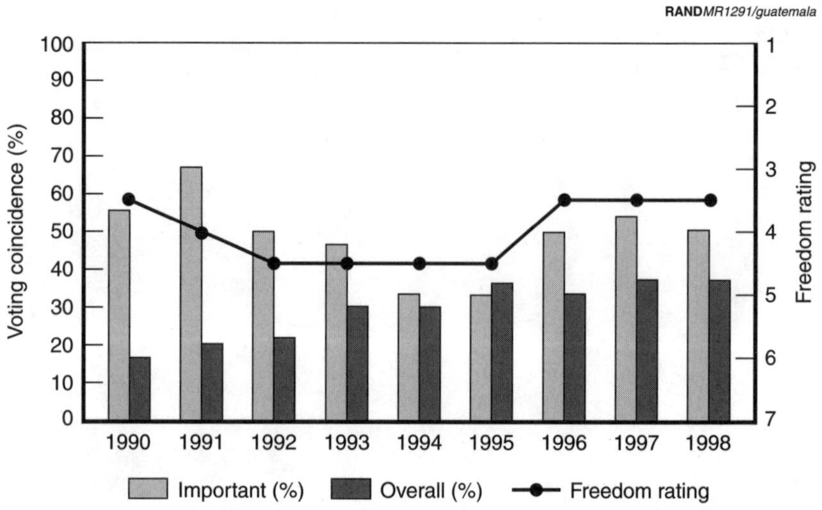

Participation in recent major operations

Desert Storm	Restore Hope	Uphold Democracy	Deliberate Force	Joint Endeavor	Allied Force	Joint Guardian
		✓				

Force size and defense spending

Ground force size	Total force size	Defense spending per soldier
38,500	40,700	approximately $2,000–$4,000

Alignment

Military:

- Rio Treaty.
- Officer training in Argentina, France, Germany, Italy, Mexico, Spain and Venezuela, U.S.

Nonmilitary:

- OAS.

- Non-Aligned Movement.

Defense and military agreements with the United States

- 1955, military assistance.
- 1954, equipment transfer.
- 1957–1974, agreements governing assistance.

Defense ties with the United States

U.S. helps train officers and senior NCOs, exercises.

Honduras

UN voting coincidence and freedom rating

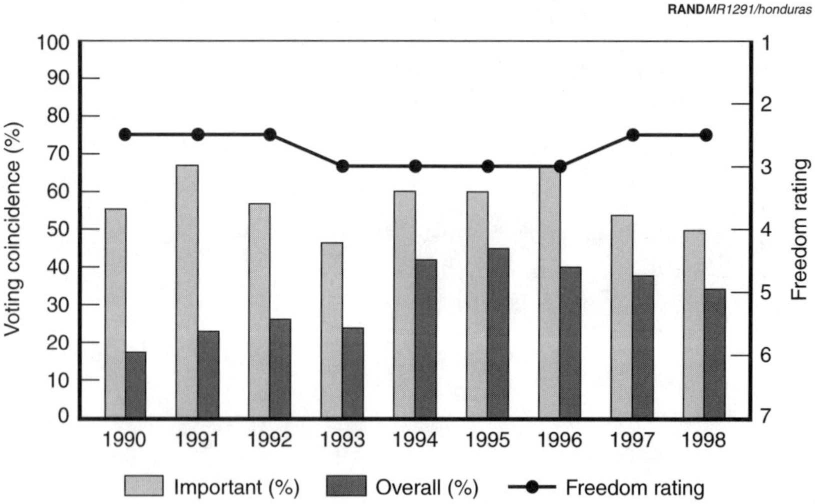

Participation in recent major operations

Desert Storm	Restore Hope	Uphold Democracy	Deliberate Force	Joint Endeavor	Allied Force	Joint Guardian
		✓				

Force size and defense spending

Ground force size	Total force size	Defense spending per soldier
16,000	18,800	approximately $1,000–$5,000

Alignment

Military:

- Rio Treaty.
- Close ties with U.S.

Nonmilitary:

- OAS.
- Non-Aligned Movement.

Defense and military agreements with the United States

- 1954, military assistance.
- 1956–1988, regulating agreements and arrangements.
- 1982, SOFA.
- 1989, Caribbean Basin Radar Network.
- 1998, provision of articles, services, and associated military education and training for anti-narcotics purposes.

Defense ties with the United States

Substantial U.S. assistance, including training; U.S. military presence, joint exercises.

Mexico

UN voting coincidence and freedom rating

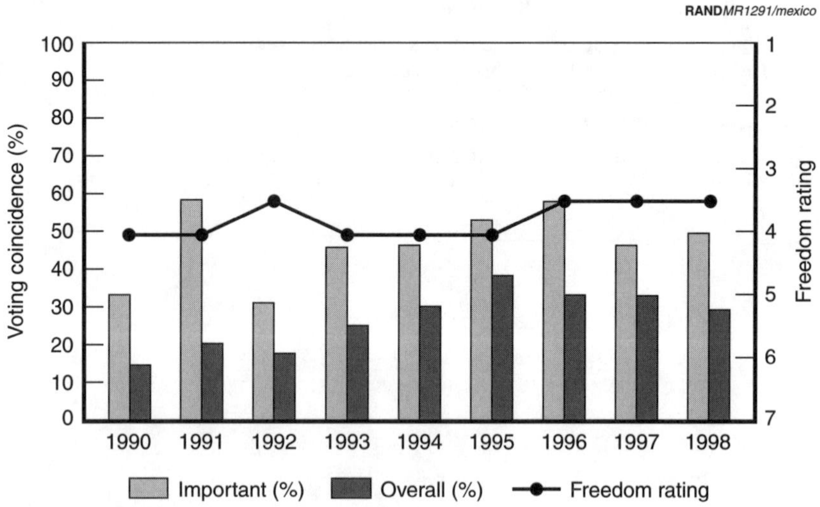

Participation in recent major operations

Desert Storm	Restore Hope	Uphold Democracy	Deliberate Force	Joint Endeavor	Allied Force	Joint Guardian

Force size and defense spending

Ground force size	Total force size	Defense spending per soldier
130,000	175,000	approximately $7,000–$21,000

Alignment

Military:

- Rio Treaty.

Nonmilitary:

- Rio Group.
- OAS.

Defense and military agreements with the United States

- 1987, IMET (1972 agreement related to training scholarships: deposit of 10% of value as military assistance).

Defense ties with the United States

No substantial defense ties.

Nicaragua

UN voting coincidence and freedom rating

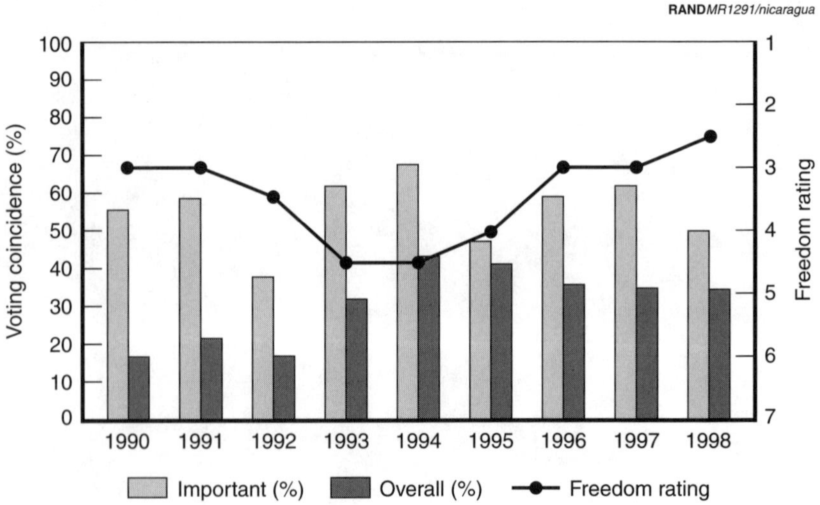

Participation in recent major operations

Desert Storm	Restore Hope	Uphold Democracy	Deliberate Force	Joint Endeavor	Allied Force	Joint Guardian

Force size and defense spending

Ground force size	Total force size	Defense spending per soldier
15,000	17,000	approximately $2,000

Alignment

Military:

- Rio Treaty.
- Training by Taiwan, others.
- To begin some U.S. training.

Nonmilitary:

- OAS.

- Non-Aligned Movement.

Defense and military agreements with the United States

- 1954, military assistance (and related through 1974).

Defense ties with the United States

Military training planned. Militaries cooperated in Hurricane Mitch clean-up.

South America (9 countries)

	Important Votes	Overall Votes	Freedom Rating	Number of Ops	Ground Strength	Per-Soldier Spending	Ties & Alignment
Argentina	0.60	0.36	2.44	1	41,000	$47,000–$64,000	High
Bolivia	0.58	0.33	2.44	0	25,000	$5,000	Medium
Brazil	0.53	0.29	3.00	0	200,000	$44,000–$47,000	Medium
Chile	0.60	0.34	2.06	0	51,000	$13,000–$23,000	Medium
Ecuador	0.57	0.32	2.56	0	50,000	$12,000	High
Paraguay	0.62	0.32	3.33	0	14,900	$7,000	Low
Peru	0.53	0.31	4.39	0	85,000	$10,000	Medium
Uruguay	0.58	0.33	1.67	0	17,600	$12,000	Low
Venezuela	0.52	0.30	2.61	0	34,000	$16,000–$17,000	Medium

NOTE: The first three columns report average values, 1990–1998.

KEY	Imp. Votes[a]	Overall Votes[a]	Freedom Rating	Number of Ops.	Ground Strength[b]	Per-Soldier Spending[c]	Alignment
High	0 to 33%	0 to 25%	1 to 3	4 to 7	>60,000	>$94,000	Frequent regularized contacts, training, agreements, with U.S.
Medium	33 to 60%	25 to 45%	3.1 to 5.5	1 to 3	20,000–60,000	$19,000–$94,000	Infrequent or irregular contacts **or** strong ties with U.S. allies
Low	60 to 100%	45 to 100%	5.5 to 7	0	5,000–19,999	<$19,000	Few, sporadic, or no contacts

[a]For UN votes, the high, medium, and low range cutoffs represent the 0–25th, 25–75th, and 75–100th percentiles, respectively.

[b]Ground-strength low, medium, and high cutoffs are the assessed ability to contribute to multinational operations at a strength of approximately one brigade, one division, or more, respectively. Based on 1997 data.

[c]Defense-spending high, medium, and low cutoffs are the percentage of U.S. per-soldier spending at 20% or less, 20–50%, and over 50%, respectively. Based on 1997 data.

Argentina

UN voting coincidence and freedom rating

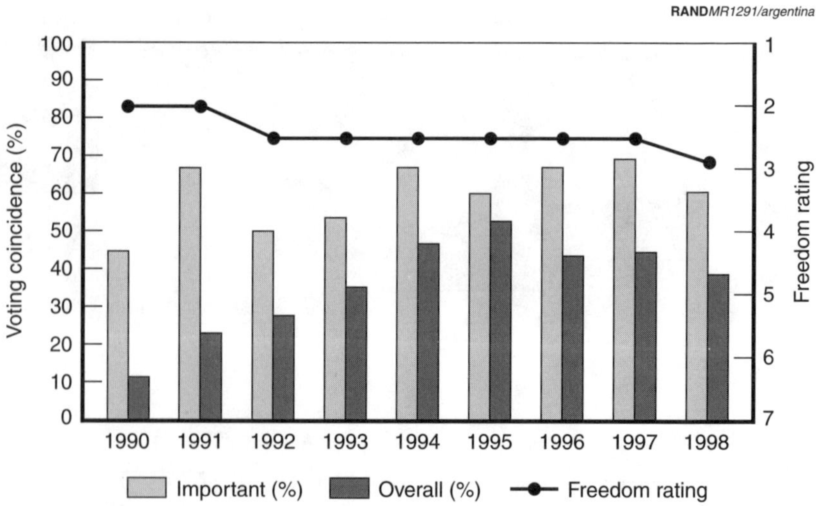

Participation in recent major operations

Desert Storm	Restore Hope	Uphold Democracy	Deliberate Force	Joint Endeavor	Allied Force	Joint Guardian
		✓				

Force size and defense spending

Ground force size	Total force size	Defense spending per soldier
41,000	73,000	approximately $47,000–$64,000

Alignment

Military:

- Rio Treaty.
- Revitalized relationship with Brazil.
- "Informal" alliance with Peru and Bolivia.
- Plan to expand military exercises with other MERCOSUR (Brazil, Paraguay, Uruguay) states (domestic in-unit training).

- CAECOPAZ peacekeeping training center has hosted personnel from Bolivia, Brazil, Ecuador, El Salvador, Guatemala, U.S.

Nonmilitary:

- Rio Group.
- OAS.

Defense and military agreements with the United States

- "Major non-NATO ally."
- 1964, military assistance program.
- 1970, armed forces cooperative projects.
- 1987, marine officer exchange.
- 1992, naval personnel exchange.

Defense ties with the United States

Exercises, training. U.S. personnel have attended CAECOPAZ peacekeeping training center in Argentina. Participates in Ecuador-Peru Military Observers Mission (Brazil, Argentina, Chile, U.S.).

Bolivia

UN voting coincidence and freedom rating

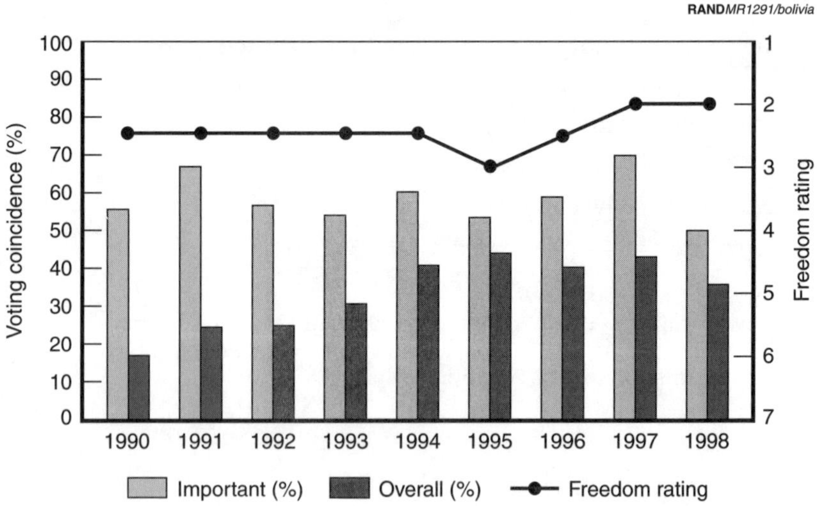

Participation in recent major operations

Desert Storm	Restore Hope	Uphold Democracy	Deliberate Force	Joint Endeavor	Allied Force	Joint Guardian

Force size and defense spending

Ground force size	Total force size	Defense spending per soldier
25,000	33,500	approximately $5,000

Alignment

Military:

- Rio Treaty.
- "Informal" alliance with Argentina and Peru.

Nonmilitary:

- Andean Community.
- Rio Group.

- OAS.
- Non-Aligned Movement.

Defense and military agreements with the United States

- 1958–1974, military assistance (initial agreement in 1958, 1962 agreement on furnishing defense articles and services; and regulating follow-ons).
- 1998, provision of articles, services, and associated military education and training for anti-narcotics purposes.

Defense ties with the United States

Majority of officers receive partial training in U.S. and the School of the Americas. U.S. training helped create battalion of special forces to fight drug trafficking, cooperating with the Army, Navy, and Air Force. Military exercises. General agreement on drug interdiction cooperation.

Brazil

UN voting coincidence and freedom rating

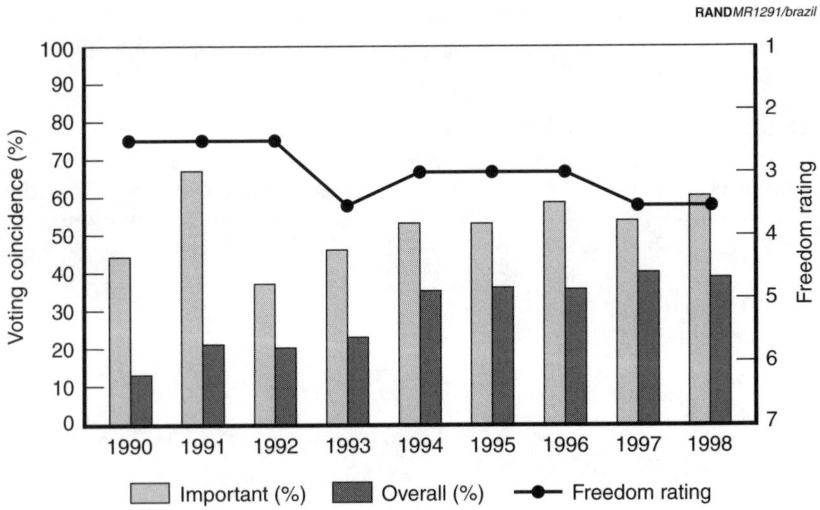

Participation in recent major operations

Desert Storm	Restore Hope	Uphold Democracy	Deliberate Force	Joint Endeavor	Allied Force	Joint Guardian

Force size and defense spending

Ground force size	Total force size	Defense spending per soldier
200,000	314,700	approximately $44,000–$47,000

Alignment

Military:

- Rio Treaty.

Nonmilitary:

- Rio Group.
- OAS.

Defense and military agreements with the United States

- 1947/48, reciprocal SOFA.
- 1953, military assistance (terminated 1977 except for safeguard clauses).
- 1964, understanding on military assistance.
- 1972, 1973, arrangements for assistance.
- 1984, industrial and military cooperation.
- 1984, data exchange for mutual development of equipment.
- 1984, (1990 extension) scientist exchange.
- 1989, IMET.

Defense ties with the United States

Current tactical and operational doctrines largely based on U.S. Participates in Ecuador-Peru Military Observers Mission (Brazil, Argentina, Chile, U.S.).

Chile

UN voting coincidence and freedom rating

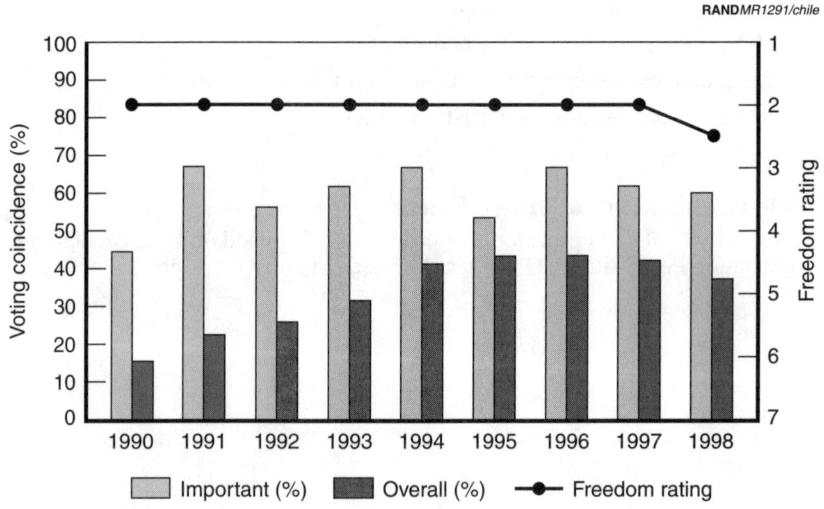

Participation in recent major operations

Desert Storm	Restore Hope	Uphold Democracy	Deliberate Force	Joint Endeavor	Allied Force	Joint Guardian

Force size and defense spending

Ground force size	Total force size	Defense spending per soldier
51,000	94,300	approximately $13,000–$23,000

Alignment

Military:

- Rio Treaty (with reservation).
- Trains some German, Israeli, Spanish, Taiwan, U.S., and other Latin American troops in military establishments.
- Army doctrine owes much to Israel.
- Considerable Israeli military assistance since 1970s.

- Recent joint exercises with Argentina and Peru as part of effort to mend fences.
- Some recent engagement with Malaysia, including defense-related memorandum.

Nonmilitary:

- Non-Aligned Movement.
- Rio Group.
- OAS.

Defense and military agreements with the United States

- 1952, military assistance (1956, 1972 governing agreements/arrangements).
- 1989, 1992, marine exchanges.
- 1992, protection of classified information.

Defense ties with the United States

Participates in Ecuador-Peru Military Observers Mission (Brazil, Argentina, Chile, U.S.). U.S. Secretary of Defense visited in 1996, exercises.

Ecuador

UN voting coincidence and freedom rating

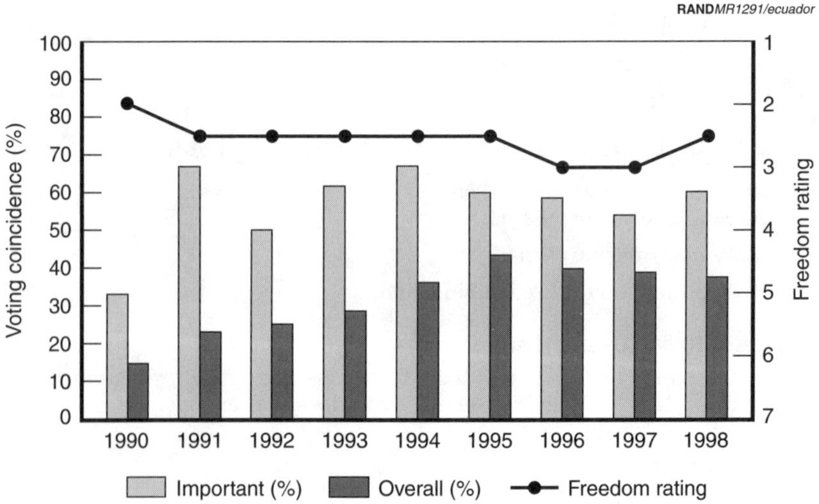

Participation in recent major operations

Desert Storm	Restore Hope	Uphold Democracy	Deliberate Force	Joint Endeavor	Allied Force	Joint Guardian

Force size and defense spending

Ground force size	Total force size	Defense spending per soldier
50,000	57,100	approximately $12,000

Alignment

Military:

- Rio Treaty.

Nonmilitary:

- Andean Community.
- Rio Group.
- OAS.

- Non-Aligned Movement.

Defense and military agreements with the United States

- 1956, return of military assistance.
- 1976, eligibility for military assistance per International Security Assistance and Arms Export Control Act.
- 1985, security of military information.
- 1992, furnishing of defense articles and services.

Defense ties with the United States

Providing facilities to make up for loss of Panama bases in drug war.

Paraguay

UN voting coincidence and freedom rating

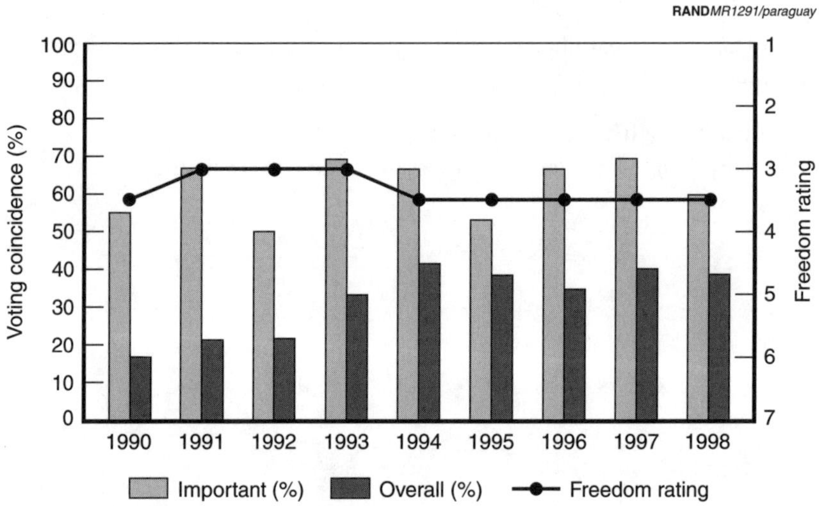

Participation in recent major operations

Desert Storm	Restore Hope	Uphold Democracy	Deliberate Force	Joint Endeavor	Allied Force	Joint Guardian

Force size and defense spending

Ground force size	Total force size	Defense spending per soldier
14,900	20,200	approximately $7,000

Alignment

Military:

- Rio Treaty.

Nonmilitary:

- Rio Group.
- OAS.

Defense and military agreements with the United States

- 1962, improving Paraguayan air transport capability.
- 1964, Army road construction and maintenance capability.
- 1966, military assistance (related 1974).
- 1995, U.S. technical personnel deployments.

Defense ties with the United States

Technical cooperation.

Peru

UN voting coincidence and freedom rating

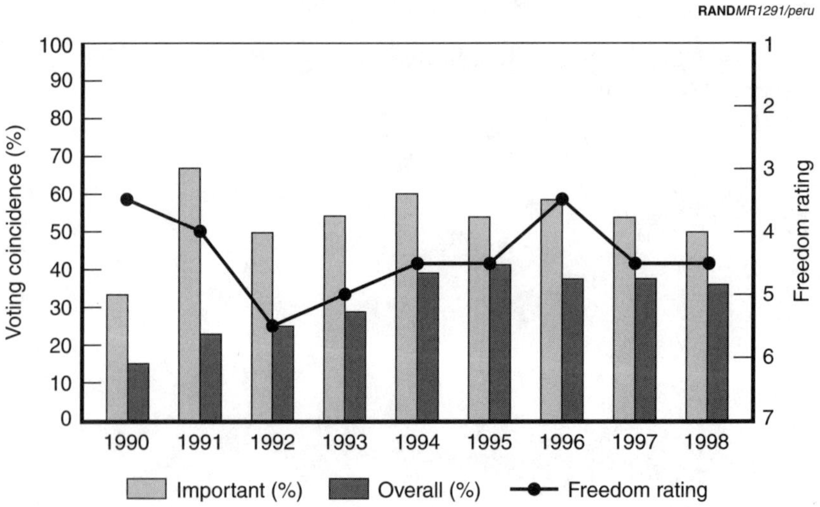

Participation in recent major operations

Desert Storm	Restore Hope	Uphold Democracy	Deliberate Force	Joint Endeavor	Allied Force	Joint Guardian

Force size and defense spending

Ground force size	Total force size	Defense spending per soldier
85,000	125,000	approximately $10,000

Alignment

Military:

- Rio Treaty.
- "Informal" alliance with Argentina and Bolivia.

Nonmilitary:

- Andean Community.
- Rio Group.

- Non-Aligned Movement.

Defense and military agreements with the United States

- 1952, military assistance (related 1955, 1962, 1972).
- 1986, officer exchange.
- 1995, SOFA (radar site and other locations as agreed).

Defense ties with the United States

U.S. radar site in country, agreement to cooperate on drug interdiction.

Uruguay

UN voting coincidence and freedom rating

RANDMR1291/uruguay

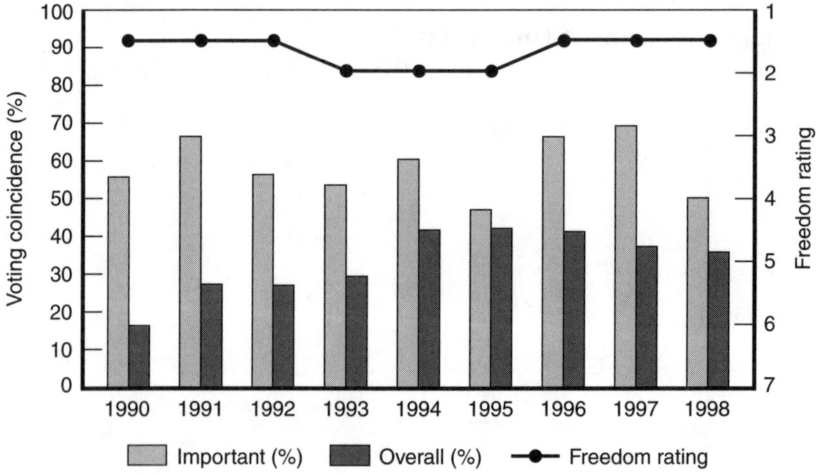

Participation in recent major operations

Desert Storm	Restore Hope	Uphold Democracy	Deliberate Force	Joint Endeavor	Allied Force	Joint Guardian

Force size and defense spending

Ground force size	Total force size	Defense spending per soldier
17,600	25,600	$12,000

Alignment

Military:

- Rio Treaty (with reservation).

Nonmilitary:

- Rio Group.
- OAS.

Defense and military agreements with the United States

- 1952, military assistance (related 1955, 1972, 1974).

Venezuela

UN voting coincidence and freedom rating

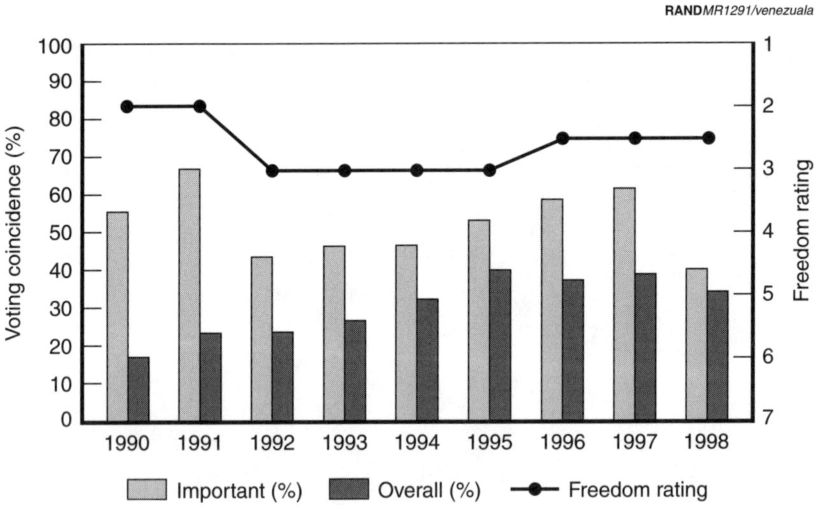

Participation in recent major operations

Desert Storm	Restore Hope	Uphold Democracy	Deliberate Force	Joint Endeavor	Allied Force	Joint Guardian

Force size and defense spending

Ground force size	Total force size	Defense spending per soldier
34,000	56,000	approximately $16,000–$17,000

Alignment

Military:

- Rio Treaty (with reservation).

Nonmilitary:

- Andean Community.
- Rio Group.
- OAS.

- Non-Aligned Movement.

Defense and military agreements with the United States

- 1976, military assistance including training (International Security Assistance and Arms Export Control Act, related 1972, 1996).
- 1983, military information security.

Defense ties with the United States

Drug interdiction cooperation, related training.

SOURCES

UN voting coincidence:

U.S. State Department, *Voting Practices in the United Nations*, Washington, D.C.: 1990–1998.

Freedom rating:

Freedom in the World, New York: Freedom House, 1990–1998.

Force size and budget figures:

International Institute for Strategic Studies, *The Military Balance 1997/98*, Oxford, UK: Oxford University Press, 1997.

International Institute for Strategic Studies, *The Military Balance 1998/99*, Oxford, UK: Oxford University Press, 1998.

Stockholm International Peace Research Institute, *SIPRI Yearbook 1998: Armaments, Disarmament and International Security*, New York: Oxford University Press, 1998.[†]

Ranges reflect variation in estimates.

Data on participation in major military operations:

Anthony H. Cordesman and Abraham R. Wagner, *The Lessons of Modern War, Volume IV: The Gulf War*, Boulder, CO: Westview Press, 1996.

Adam B. Siegel, *The Intervasion of Haiti*, Alexandria, VA: Center for Naval Analyses, Professional Paper 539, August 1996.

[†]1997 defense budgets were reported in this source in 1995 U.S. dollars. When used here, they have been adjusted to 1997 U.S. dollars with the inflation factor reported for that two-year period in this source: 5.4 percent.

Larry Wentz (ed.), *Lessons from Bosnia: The IFOR Experience*, Washington, D.C.: NDU Press, 1998.

Data on operations Restore Hope, Deliberate Force, and Allied Force were obtained at the Federation of American Scientists (FAS) Web site, *www.fas.org/man/dod-101/ops/index.html,* and at the NATO AFSOUTH Web site, *www.afsouth.nato.int.*

Data on participation in Operation Joint Guardian is based on KFOR Press Information Center, letter to author, September 23, 1999; "NATO Update 9–15 June 1999," *www.nato.int/docu/update/1999/u990609e.htm;* and "Speaking Notes by the Chairman of the Military Committee at his Initial Press Conference," *www.nato.int/docu/speech/1999/s990630a.htm;* and U.S. and international press coverage.

Data on alignment and relations:

U.S. Department of State, *Treaties in Force*, Washington, D.C., 1997.

U.S. Department of State, "Current Treaty Actions":

www.state.gov/www/global/legal_affairs/treaty_actions_1999.html
www.state.gov/www/global/legal_affairs/treaty_actions_1998.html
www.state.gov/www/global/legal_affairs/treaty_actions.html.

Charles Heyman (ed.), *Jane's World Armies*, Coulsdon, Surrey, UK, Alexandria, VA: Jane's Information Group, 1999.

U.S. and international press coverage.

Appendix B

MILITARY COMPATIBILITY ASSESSMENT TOOL

In 1998 RAND designed the DynaRank decision-support system, a Microsoft Excel workbook available for Macintosh and Windows-based computers, to assist Department of Defense decisionmaking. DynaRank ranks policy options by cost-effectiveness based on the relative importance of objectives and a variety of success criteria. Although DynaRank aimed to assist the DoD's high-level resource allocation decisionmaking, its strong sensitivity to strategy, amenability to a variety of data (subjective judgments as well as quantitative analyses), and ability to link several levels of analysis make it useful in supporting other types of defense planning.[1]

By substantively modifying the DynaRank system, we developed the Military Compatibility Assessment Tool (MCAT). The tool aims to assist DUSA-IA in making strategic-level MFC policy decisions. MCAT does so by pinpointing the extent of compatibility between U.S. ground forces and other countries' armies in specified critical areas, thereby providing a rationale for MFC policy choices. Linking specific MFC assistance policies with the results from MCAT also aids the decisionmaking process for allocating resources for MFC activities.

MCAT analysis further differentiates between the countries identified as likely to participate with the United States in coalition operations (on the basis of the propensity to ally). It provides a means of evalu-

[1]For further details on DynaRank, refer to Richard J. Hillestad and Paul K. Davis, *Resource Allocation for the New Defense Strategy: The DynaRank Decision-Support System*, Santa Monica, CA: RAND, 1998.

ating the existing level of military compatibility of select foreign ground forces relative to the U.S. Army in a variety of critical mission areas. In addition, MCAT allows for variable weighting of requirements based on different missions, making the tool flexible and adaptable across the mission spectrum. An MCAT analysis produces a hierarchy of countries based on their ground forces' compatibility with the U.S. Army and identifies both areas of greatest need and potential bottlenecks in coalition operations with the United States. It does so in a relative fashion, allowing for comparisons among the countries while judging all against the United States as the standard point of reference. The hierarchy provides input to assist DUSA-IA in making decisions about which countries and which capability assessment areas (CAAs) need attention if they are to operate more effectively in coalition operations with the United States. MCAT helps decisionmakers pinpoint the most effective policies for enhancing MFC.

METHODOLOGY

The MCAT provides a method for general assessment of a foreign army's current capabilities and associated compatibility with U.S. ground forces. This flexible tool enhances the user's ability to evaluate multiple countries based on a variety of criteria; change criterion weights according to mission requirements; and consolidate the evaluations for comparison.

Initially, the user conducts an analysis of a country's capabilities and provides an assessment (score) of the related compatibility in the designated CAAs. Once the country assessments are complete, the user opens the MCAT workbook, which consists of three worksheets: The country data worksheet; the MFC scorecard; and a scorecard functions explanation sheet. The user enters the scores onto the country data worksheet, which automatically transfer to the MFC scorecard. Utilizing the MFC scorecard, the user sets CAA weights based on mission requirements, and colors the scorecard. In the final step, the user determines the type of analysis by appropriately flagging the desired compatibility measures (CMs) for each country and then ranking the countries by effectiveness. The effectiveness ranking creates a hierarchical listing of the countries by "overall" compatibility or compatibility in any one of the three CMs.

The quality of the MCAT results depends upon the quality of the input information. The underlying country evaluations must be both credible and understandable. The evaluations, while primarily subjective, may come from specific detailed analysis, spreadsheet models or other quantitative methods, or assessments based on general and specific knowledge and experience. The sensitive nature of the information required to determine a country's military capability and compatibility with U.S. ground forces makes consolidated data in unclassified form difficult to obtain. However, the Military Capabilities Studies, conducted for specific countries by the Defense Intelligence Agency, supply current and accurate analysis of a country's capabilities from which to make assessments. Additional sources of information include the plans and intelligence sections of Joint, CINC, and Army staffs, CIA country studies, military attachés at U.S. embassies, and open sources such as Jane's publications.

Basic Structure of the MFC Scorecard

The MCAT, an Excel-based tool, consists of two worksheets. The user enters data on the base-level worksheet for each country in nine capability assessment areas (maneuver, fire support, air defense, mobility/survivability, logistics, deployability, command and control, communications, and intelligence) evaluated by three compatibility measures (doctrine, technology, and operational readiness/training). The data automatically transfer to the MFC scorecard, which presents the aggregate results of the assessment. Figure B.1 illustrates a completed MFC scorecard.[2] The scorecard illustrates both individual and aggregate levels of information. First, the "overall row" provides a country's overall level of compatibility in each CAA. Additionally, the aggregate column to the far right of the scorecard provides each country's aggregate compatibility for each CM.

This composite scorecard also provides a ranking of countries based on the compatibility assessment. Through a simple process of reweighting the flags located at the left of the first column, MCAT creates a hierarchical country ranking reflecting "overall compatibil-

---

[2]Although all illustrations in this report are rendered in black and white, the MCAT itself utilizes a full color scheme.

RANDMR1291-B.1

Color Blank Cells
3 High Color Value
0 Low Color Value

COUNTRY	CM	Flag	BATTLEFIELD FUNCTIONS				CSS		C4I			Wt or Min → Wt
		CAAs	Maneuver	Fire Support	Air Defense	Mob/Surv	Logistics	Deployability	C2	Commo	Intel	Compatibility Score Column
			1	1	1	1	1	1	1	1	1	
Country 1	Doctrine	1	3	3	3	2	2	2	3	2	3	2.6
Country 1	Technology	1	2	2	2	2	1	2	2	2	3	2.0
Country 1	OR/TRNG	1	2	2	2	2	1	1	2	2	2	1.8
Country 1	**Overall**	1	2.3	2.3	2.3	2.0	1.3	1.7	2.3	2.0	2.7	2.1
Country 2	Doctrine	1	2	2	1	1	2	1	2	2	1	1.6
Country 2	Technology	1	2	1	1	1	1	0	2	1	1	1.1
Country 2	OR/TRNG	1	2	1	1	1	1	0	2	1	1	1.1
Country 2	**Overall**	1	2.0	1.3	1.0	1.0	1.3	0.3	2.0	1.3	1.0	1.3
Country 3	Doctrine	1	1	1	1	1	2	2	1	2	1	1.3
Country 3	Technology	1	2	2	1	2	1	1	1	2	2	1.6
Country 3	OR/TRNG	1	1	1	1	1	1	1	1	1	1	1.0
Country 3	**Overall**	1	1.3	1.3	1.0	1.3	1.3	1.3	1.0	1.7	1.3	1.3

Figure B.1—Completed and Unweighted MFC Scorecard

ity" or compatibility by CM. Additionally, the scorecard allows a comparison of countries, not only by overall level of compatibility, or CM, but also by levels of compatibility within the individual capability assessment areas.

Capability Assessment Areas (Columns) / Compatibility Measures (Rows)

We derived the MCAT criteria, both CAAs and CMs, by examining multiple reports on past multinational operations and a variety of Army publications.[3] While these criteria provide a valid and critical

---

[3]Publications include Army Field Manual FM 100-5 *Operations,* June 1993; Army Field Manual FM 100-8, *The Army in Multinational Operations,* November 1997; Allied Joint Publication AJP-01(A), *Allied Joint Doctrine,* March 1999; *Joint Task Force*

examination of a country's capability/compatibility, the strength of the MCAT lies in its ability to let the analyst change the criteria if an examination of the underlying assumptions proves they are no longer valid or require adjustment as the situation may dictate.

Utility Values/Body of the MFC Scorecard

MCAT does not automatically complete the large analytic task of filling out the body of the scorecard. The user must make a subjective assessment of a country's capabilities and then relate the assessment to a compatibility score.

The underlying assumption for the country analysis is that a country's capabilities directly relate to its compatibility with U.S. ground forces and therefore provide the best method for evaluation. Differences in nations' military capabilities lead to asymmetry between forces. This asymmetry, whether doctrinal, technological, or procedural, will complicate interoperability and hinder coordination. In evaluating these asymmetries, we classify countries in terms of compatibility.

While similar capabilities often signal an increased level of compatibility between countries, we recognize that this is not always the case. Equality of capabilities between nations does not necessarily equate to compatibility between nations. For example, two countries may have the same type of equipment but use it for different missions.[4] In-depth analysis of this type of asymmetry is important when making detailed decisions about individual nations. But when evaluating relative compatibility to make strategic decisions about

---

Commander's Handbook for Peace Operations, Joint Warfighting Center, February 1995; Roger H. Paulin, *Multinational Military Forces: Problems and Prospects,* London: Institute for Strategic Studies, Adelphi Paper 294, 1995; Thomas J. Marshall (ed.), with Phillip Kaiser and Jon Kessmeire, *Problems and Solutions in Future Coalition Operations,* Carlisle Barracks, PA: U.S. Army War College, Strategic Studies Institute, December 1997; Martha Maurer, *Coalition Command and Control,* Washington, D.C.: National Defense University, Institute for National Strategic Studies and the Center for Advanced Concepts and Technology, 1994; Michele Zanini and Jennifer Morrison Taw, *The Army and Multinational Force Compatibility,* Santa Monica, CA: RAND, MR-1154-A, 2000; and Elwyn Harris, Stephanie Cammarata, Jody Jacobs, Lewis Jamison, Iris Kameny, and Paul Steinberg, unpublished RAND research, 1998.

[4]The concept of asymmetry is taken from Chapter 3 (by Steven Metz) of *Problems and Solutions in Future Coalition Operations,* op. cit.

MFC policies, the concept can be simplified, allowing for cross-comparisons between nations. We assume that a foreign army with more advanced capabilities in a particular CAA has a higher level of compatibility with U.S. ground forces relative to a country with little or no capability in that CAA. Countries with similar equipment and doctrine will have a higher compatibility level than countries with dissimilar equipment or doctrine. This assumption allows us to make subjective judgments on compatibility and cross-comparisons between nations.

The aforementioned reasoning led to the development of four levels of compatibility. The compatibility measures range from "No Capability—0" to "High Capability—3." Figure B.2 presents both the levels of compatibility and the weighting scheme used on the MFC scorecard.[5]

Note that the definitions for levels 1 and 2 (light and dark gray) derive from the ability to increase compatibility either through long-term fixes (light gray) or short-term fixes and workarounds (dark gray). A 2000 RAND study on MFC, entitled *The Army and Multinational Force Compatibility,* derived the following definitions to explain long- and short-term fixes. Broadly defined, short-term fixes or workarounds represent those steps taken prior to a deployment, once it becomes a distinct possibility (i.e., during the planning phases of an operation). Long-term fixes equal long-term solutions sustained across operations.

0	No capability (no compatibility)
1	Low capability (long-term fixes required for compatibility)
2	Medium capability (short-term fixes and workarounds will ensure compatibility)
3	High capability (highly compatible/mission ready)

Figure B.2—Compatibility Levels

[5]Although all illustrations in this report are rendered in black and white, the MCAT itself utilizes a full color scheme.

Weighting of Capability Assessment Areas

MCAT allows for varying the relative weights of the CAAs. In its default setting, MCAT places a weight factor of 1 on each CAA. Equal weights on all categories generate a cumulative score for each row that is a simple average of each individual value. We would recommend starting the analysis with the default setting, since this establishes a useful baseline estimate of how the countries compare across a number of factors.

The weights can be changed when the user is considering scenarios in which it is clear that a particular CAA (or set of CAAs) is either more or less relevant than others. For instance, if countries are assessed on their compatibility for operations with long lead times (such as another Gulf War, when the United States and its allies had several months to deploy), one could discount the importance of prompt deployability. At the extreme, the deployability CAA could be given a score of zero, implying that each country's score in this area would not count at all in the computation of the row compatibility score. Dropping deployability from the overall score will increase the average score for countries scoring low in this CAA (the opposite applies to countries with a high deployability score).[6] Other scenario variables that are likely to change the weighting scheme would include the intensity of conflict, command arrangements, and degree of involvement the United States would wish to have.

While weighting of CAAs provides a useful method of evaluating MFC across different mission types, it is subjective, and the user must understand the limitations of reweighting the CAAs. A host of results can be obtained by manipulating the weights. Therefore, adjusting the weights of the CAAs is more apt to facilitate the finding of robust options among the many different alternatives for MFC funding than to ensure that the "correct" options are highlighted.

---

[6]The effect is likely to be small if only one CAA is completely discounted. This is due to the fact that the original average score includes eight other CAAs and is therefore relatively insensitive to the removal of one out of nine inputs.

Table B.1

Framework for Deriving Mitigation Measures

Problems		Ad Hoc, High or Low Intensity, Long Lead Time
C4I	Operational	Provide C4I, liaisons; **IMET,** predeployment MTTs; **develop intel-sharing protocols**
	Organizational	Establish lead nation C2 structure, geographic separation; set up C3IC
	Technological	Loan/share/**sell** equipment; rely on lowest common denominator (LDC), COTS, SATCOM where not compromised
Logistics and Deployability	Operational	Phase deployment; provide logistics & lift; **preposition** materiel; lease lift, local transportation
	Organizational	Establish geographic separation; stovepiping
	Technological	Loan/share/sell equipment
Doctrine, Procedures and Employment	Operational	Provide liaisons; **IMET;** predeployment MTTs, **standardized** and predeployment exercises; **invite LNOs to TRADOC, War College, other Army centers;** provide missing capabilities (force protection); establish a quick reaction force
	Organizational	Establish lead nation C2 structure, geographic separation
	Technological	Loan/share/**sell** equipment; rely on COTS

*The fixes are shown in boldface type. The relative importance of fixes and workarounds changes with the amount of lead time: with short lead times, fixes become more important since some workarounds may not be feasible.

Table B.1—extended

Ad Hoc, High or Low Intensity, Short Lead Time*	Alliance, High or Low Intensity, Long or Short Lead Time
Provide C4I, liaisons; **develop intel-sharing protocols**	Provide C4I, liaison; **develop combined exercise training & intel-sharing protocols**
Establish lead nation C2 structure, geographic separation; set up C3IC	**Integrate C2 structure, forces;** partly rely on geographic separation
Loan/share/**sell** equipment; rely on LDC, COTS, SATCOM where not compromised	Loan/share/**sell** equipment; rely on LDC, **jointly develop equipment;** rely on COTS, SATCOM where not compromised
Phase deployment; provide logistics & lift; **preposition materiel;** lease local transportation	**Implement combined total asset visibility (TAV);** provide logistics & lift; **preposition materiel;** lease lift (if long lead time), local transport
Establish geographic separation; stovepiping	**Develop combined, complementary lift and logistics procedures;** or stovepipe
Loan/share/**sell** equipment	**Share, co-develop TAV; coordinate procurement to ensure compatibility**
Provide liaisons; **IMET; standardized exercises; invite LNOs to TRADOC, War College, other Army centers;** provide missing capabilities (force protection); establish a quick reaction force	**Develop combined doctrine, training, exercises, exchanges, etc.;** provide missing capabilities (force protection); and **compensate in combined planning**
Establish lead nation C2 structure, geographic separation	**Integrate command structure, forces;** partly rely on geographic separation
Loan/share/**sell** equipment; rely on COTS	Loan/share/**sell** equipment; **co-develop equipment and materiel; establish compatibility protocols**

Flagging of Compatibility Measures

To facilitate different levels of analysis, the user may employ the flagging function to determine the type of hierarchy MCAT produces. Based on the flagging, MCAT produces country rankings either by overall compatibility or by any one of the three CMs (doctrine, technology, or operational readiness). Figure C.3 in the next appendix provides an example of flagging the CMs for analysis.

Linking MCAT Data to MFC Assistance Policies

The previously mentioned 2000 RAND study of MFC analyzed the effects of Force XXI on MFC in future coalition operations. A close study of four major multinational operations found repeated use of certain "mitigation measures" to overcome compatibility problems between the United States and its coalition partners. Table B.1 outlines the results of the study's findings.

The study also demonstrated the value of engagement over the long term, or the advantage of fixes over workarounds. Such initiatives as combined training, multilateral command post exercises, technological research and development, doctrinal development, and intelligence-sharing protocols will be increasingly important as forces with vastly different capabilities attempt to coordinate their efforts.[7]

Linking the MCAT results to this study reveals several significant policy applications. MCAT analyzes a country's compatibility with U.S. ground forces relative to other possible coalition partners and identifies shortfalls in compatibility, categorizing them by both capability area and compatibility measure. Not only can the user identify a compatibility shortfall by its functional area, (e.g., intelligence, C2), he can further identify whether the problem is a doctrinal, technological, or operational readiness shortfall. The Zanini and Taw study provides a sample of the types of mitigation measures available to the Army to address compatibility.

---

[7]Zanini and Taw.

Appendix C

MCAT TUTORIAL

APPLICATION OF MCAT FOR MFC STRATEGIC PLANNING

The following steps outline how to use the MCAT. The remainder of the section gives further details for each step.

Step 1: Fill out country data worksheet.

Step 2: Transfer data to MFC scorecard.

Step 3: Set weights for CAAs on MFC scorecard.

Step 4: Color MFC scorecard.

Step 5: Set flags for type of analysis and rank MFC scorecard for analysis.

Note that steps displayed below in boxes denote computer steps.

STEP 1: Fill Out the Country Data worksheet

- Open the MCAT workbook.
- Go to "country data worksheet."
- Enter the country name over "Country 1." "Return" will replace all appropriate areas with the country name.
- Enter country compatibility values into appropriate rows/columns. **Do not** input data into overall column.

Country data worksheet. The worksheet body contains the information to be entered on the scorecard at the intersection of the CAAs (columns) and the CMs (rows). The user fills out this portion of the worksheet using primarily subjective assessments of each country based on analytical research of its capabilities and their associated compatibility with U.S. ground forces. (Figure B.2 in the previous appendix illustrates the four levels of compatibility used for evaluation.)

Compatibility measures. The user evaluates each CAA using three CMs. These CMs, drawn from a review of recent multiforce operation after-action-reports and studies and current Army doctrine and regulations, represent the critical measures affecting the compatibility between two forces.[1] While compatibility problems are often attributed to equipment or technology differences between nations, MCAT takes into account the more complicated and realistic view that compatibility is based on a variety of factors. The compatibility measures used in the MCAT are doctrine, technology, and operational readiness/training.

- **Doctrine.** How do they fight? Does their doctrine cover the full treatment of strategic, operational, and tactical issues? Does it cover Joint and Combined operations? Does it emphasize offense, defense, or both? What type of operations does it cover?
- **Technology.** What do they fight with? What is their modernization level?
- **Operational readiness/training.** What are their forces capable of? What types of missions are they trained to conduct? What do recent evaluations/exercises demonstrate about the readiness of their forces?

Capability assessment areas. The 2000 RAND study on MFC issues categorized the compatibility issues identified in each of the four operations analyzed into three specific areas: C4I, logistics and deployability, and doctrine, procedures, and employment.[2] To enhance the evaluation, we further divided them into nine compatibil-

[1]Publications reviewed include FM 100-5, FM 100-8, Paulin, Marshall, and Maurer.

[2]Zanini and Taw.

ity assessment areas derived from both the above-mentioned study and Army doctrine on multinational operations.[3] The following sets of questions for each of the nine CAAs are designed to guide the user in the capability assessment.[4]

1. Maneuver

- **Doctrine.** How do they fight? Is their doctrine Soviet-based or U.S.-based? Primarily defensive or offensive oriented? What is the size of their basic maneuver force? Do they plan for combined arms operations, joint operations, combined operations?
- **Technology.** What is the modernization level of their maneuver equipment? What is the mobility of their formations? What type of communications equipment do their maneuver formations have?
- **OR/Training.** What is the effectiveness of their formations? What is their assessed maneuver capability? Do they have training for operating in special environments (airborne, air-assault, urban, NBC)?

2. Fires

- **Doctrine.** What is their doctrine on allocation of firepower? Who controls their firepower assets? What are their fire control measures?
- **Technology.** What type of targeting/firepower equipment do they have?
- **OR/training.** What is their capability for long-range/massed fires? What is their capability to integrate and synchronize fires?

---

[3]FM 100-5, FM 100-8.

[4]The list of questions was drawn from analysis of FM 100-5 and FM 100-8 and most likely does not include all relevant ones. However, including additional questions may unnecessarily complicate the analysis. There are multitudes of questions pertaining to each CAA, and delineation of every relevant one would be overwhelming and unproductive. To complete a general, relative assessment of compatibility across nations with the MCAT, a basic subjective assessment of a nation's capability is required.

3. Air Defense

- **Doctrine.** What are their priorities for air defense? Who controls their AD assets? What are their fire control measures?
- **Technology.** What are their weapons systems types and capabilities? What type of communications equipment do they have? Do they have theater missile defense capability?
- **OR/training.** What are their AD capabilities? Can they integrate their fires into the theater control and reporting center (CRC)?

4. Mobility/Survivability/Force Protection

- **Doctrine.** Does it cover the use of mobility/countermobility measures? Do they plan for force protection?
- **Technology.** What type of engineer equipment do they have? What assets do they maintain for force protection? What type of NBC equipment do they have?
- **OR/training.** What type of mobility/countermobility capabilities do they have? Can they provide survivability measures for their own forces, particularly NBC?

5. Logistics

- **Doctrine.** Do they have a push or a pull logistics doctrine? Do they plan for RSO&I operations? What classes of supply can they provide, and what are the priority and stockage levels? What types of services do they plan on providing? What are their maintenance standards? How long do they plan on sustaining initial forces? What are their plans for follow-on sustainment of forces?
- **Technology.** What is their modernization level for equipment? What is their logistics mobility capability? What type of system (manual/ADP) do they utilize to track their logistics?
- **OR/training.** What size of a deployed force can they sustain, with what classes of supply and services and for how long? What level of maintenance do they achieve?

6. Deployability

- **Doctrine.** Does their doctrine cover strategic deployment of forces? Do they plan for RSO&I activities?
- **Technology.** What modes of transport do they maintain (land, air, sea, rail)? What technology do they utilize for planning deployment (manual/ADP)?
- **OR/training.** What type of deployment can they conduct? What size force can they self-deploy, to what distance, and at what speed?

7. Command and Control

- **Doctrine.** Do they have large staffs with the technical means to support planning (command estimate process) for both current and future operations? Do they maintain battle staffs? Can they process, reproduce, and rapidly disseminate operational plans? What is the decisionmaking authority of the subordinate commanders and staffs (decentralized or centralized decisionmaking processes)? Do they have a flexible command structure giving commanders the freedom to execute the mission with minimal guidance? What is their reporting system? What graphics and control measures do they utilize to control the battlefield?
- **Technology.** Do they have the communications, computers and intelligence means to support the commanders' decisionmaking process? Do they have the ability to use space-based systems for reconnaissance, surveillance, navigation, and positioning to facilitate battle command?
- **OR/training.** What is the assessment of their C2 structure from current exercises and operations?

8. Communications

- **Doctrine.** What type of orders format and dissemination methods do they utilize? Do they plan for communications capabilities for home station, en-route, and in-theater operations? Can they operate using English, or do they plan for liaisons and translators?

- **Technology.** What types of signal support systems are available to the commanders at different levels? Are their systems digital or analog? Do they allow for digital-to-analog interfaces? Do they have secure communications capabilities and can they tie into U.S. systems (STUs, SINCGARS, SATCOM)? Do they have access to space-based systems?
- **OR/training.** What is the assessed level of their communications capabilities? Are their communications systems reliable and can they establish commo channels in a timely manner? What is their level of "U.S. military English" proficiency (officers, NCOs, enlisted)? Do they have liaisons with good English language skills and understanding of U.S. doctrine and procedures? What are their translation capabilities?

9. Intelligence

- **Doctrine.** What type of collection capabilities/assets do they possess (HUMINT, SIGINT, IMINT, ELINT), and how do they utilize them in support of political and military objectives?
- **Technology.** What is the sophistication of their collection assets? What technical means do they have to produce and disseminate intelligence? Can they link into the U.S. intelligence system (i.e., SATCOM)?
- **OR/training.** What is their capability (sophistication and focus) for processing intelligence (collection, production, and dissemination)?

The remaining four steps describe how to transfer country data to the scorecard and configure it for analysis. Figure C.1 demonstrates the correct data-entry points for both weights and flags.

STEP 2: Transfer data to scorecard

- Select "MFC scorecard" worksheet from MCAT workbook. Data from country data worksheet automatically transfer to the scorecard.

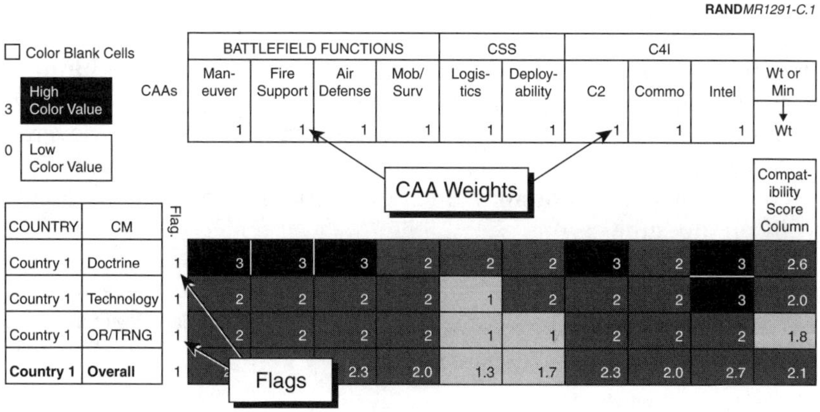

Figure C.1—Partial MFC Scorecard Demonstrating Weight and Flag Options

STEP 3: Set CAA weights on MFC scorecard

- Set weights for CAAs on scorecard, typing the numbers in the correct columns.

STEP 4: Color scorecard

The user should beware that the coloring in MCAT is not done automatically on a scorecard, owing to software limitations. Changes made to weights and country values require the colors to be reset. To remove color, repeat the first two steps and select "Uncolor Scorecard."[5]

- Go to Microsoft Excel toolbar menus.
- Select "SC Tools" and select "Color Scorecard."

---

[5]Hillestad and Davis, *Resource Allocation for the New Defense Strategy*, 1998.

STEP 5: Set flags for type of analysis and rank MFC scorecard

To rank the countries, place the same number in the flag column of all of the options you want ranked together, and place a different, lower number in the remaining rows. For example, to rank by overall category, place a 1 in the overall category for each country and a 0 in all other rows. It is possible to rank all similar categories hierarchically on the same card.

- Set flags for type of analysis, typing the number in the appropriate flag column.
- Go to Microsoft Excel toolbar menus.
- Select "SC Tools."
- Select "Rank."
- Select "By Effectiveness" (**do not** select "By Cost" or "By Cost Effectiveness").[6]

SAMPLE MCAT WORKSHEET AND SCORECARD

In the following example, we apply the MCAT process to three countries, providing specific capability/compatibility analysis for Country 1 and sample scores for Countries 2 and 3. Figure C.2 provides an example of a completed country data worksheet.

Country 1 Assessment

1. Maneuver

- **Doctrine.** Doctrine covers full spectrum of combined arms, joint, and combined operations. Country 1 has traditionally emphasized defensive operations, although the latest national military strategy emphasizes offensive, power-projection missions. Its ground forces are centered around the brigade-sized maneuver unit, and they follow U.S. doctrine.

  Score: 3

[6]These functions are a DynaRank option not adapted to MCAT analysis.

RANDMR1291-C.2

		Maneuver	Fire Support	Air Defense	Mob/Surv	Logistics	Deploy-ability	C2	Commo	Intel
Country 1	Doctrine	3	3	3	2	2	2	3	2	3
Country 1	Technology	2	2	2	2	1	2	2	2	3
Country 1	OR/TRNG	2	2	2	2	1	1	2	2	2
Country 1	Overall	2.3	2.3	2.3	2.0	1.3	1.7	2.3	2.0	2.7
Country 2	Doctrine	2	2	1	1	2	1	2	2	1
Country 2	Technology	2	1	1	1	1	0	2	1	1
Country 2	OR/TRNG	2	1	1	1	1	0	2	1	1
Country 2	Overall	2.0	1.3	1.0	1.0	1.3	0.3	2.0	1.3	1.0
Country 3	Doctrine	1	1	1	1	2	2	1	2	1
Country 3	Technology	2	2	1	2	1	1	1	2	2
Country 3	OR/TRNG	1	1	1	1	1	1	1	1	1
Country 3	Overall	1.3	1.3	1.0	1.3	1.3	1.3	1.0	1.7	1.3

Figure C.2—Sample Excel Spreadsheet

- **Technology.** Country 1's maneuver equipment is comparable with U.S. standards for hardware. However, it has not completed recent "software" upgrades in communications and targeting.

  Score: 2

- **OR/training.** Maneuver forces are capable of conducting fast-paced operations in different terrain. However, Country 1's forces are not suited for operating in a WMD environment—they lack adequate NBC equipment and protection and decontamination capabilities.

  Score: 2

2. Fires

- **Doctrine.** Plans for wide use of long-range massed fires, both organic and joint, to support both offensive and defensive operations. Maneuver-unit (brigade) commander has control over tactical firepower assets.

  Score: 3

- **Technology.** A recent ATACMS acquisition program will significantly increase Country 1's technological capability to fight deep. It uses both self-propelled and towed artillery compatible with U.S. standards. Country 1's artillery units have effective fire control centers. However, due to lack of upgrades in software, these units are limited in their real-time targeting capabilities.

  Score: 2

- **OR/training.** Country 1 is effective in controlling fires for the close battle. Its ability to coordinate joint fires remains less than adequate, and organic deep assets have yet to be integrated.

  Score: 2

3. Air Defense

- **Doctrine.** Country 1's doctrine develops a comprehensive set of fire control measures, compatible with U.S. standards.

  Score: 3

- **Technology.** No deployable theater missile defense (TMD) system. Country 1 possesses organic air defense (AD) assets, requiring laser-target designators. Identification friend or foe (IFF) based primarily on visual recognition.

  Score: 2

- **OR/training.** Limited IFF capability hampers Country 1's training and operational readiness.

  Score: 2

4. Mobility/Survivability/Force Protection

- **Doctrine.** Doctrine covers full spectrum of mobility/countermobility, but force protection is less developed.

  Score: 2

- **Technology.** Substantial and modern engineering equipment with high mobility. Inadequate NBC protection and decontamination equipment, as well as lack of TMD, weakens survivability.

  Score: 2

- **OR/training.** Fully capable of conducting engineer activities across the battlefield. However, lack of adequate force protection and NBC training severely decreases Country 1's readiness in this area.

  Score: 2

5. Logistics

- **Doctrine.** "Pull" logistics doctrine is a legacy from this country's traditionally defensive orientation. Recent reforms in doctrine place increased emphasis on a "push" and joint approach. Country 1 can provide the full range of supply classes and required services to sustain its force for regional operations. Doctrine sets high maintenance standards.

  Score: 2

- **Technology.** Despite recent acquisitions and equipment upgrades to enhance deployability, Country 1 continues to suffer from low mobility of its logistics assets. It has not fully developed a capability to support rapid offensive operations or strategic deployments and continues to rely primarily on manual tracking system for logistics.

  Score: 1

- **OR/training.** Limited capabilities to support strategic deployment. The logistics units of Country 1 have not reconfigured fully to provide both robust and economical organic support to the combat forces and sound higher-echelon support to reinforce, replenish, and resupply the organic capability. Cutbacks in defense spending have led to a decrease in overall maintenance standards, and low stockage levels of repair parts.

  Score: 1

6. Deployability

- **Doctrine.** Country 1 began to develop doctrine on strategic deployments in 1995. Current doctrine calls for rapid force projection and the ability to conduct protracted operations abroad.

  Score: 2

- **Technology.** It has pursued a vigorous program to reorganize and equip its forces to meet doctrinal requirements for force projection. Recent acquisition programs have given Country 1 the technological capability to rapidly self-deploy a brigade-size force in support of strategic operations anywhere in the world.

  Score: 2

- **OR/training.** While Country 1 is doctrinally and technologically prepared to self-deploy for operations abroad, its forces require significant training and self-deployment functions such as strategic lift planning and RSO&I. The limited ability to provide critical supply functions hampers Country 1's ability to sustain forces deployed outside of the region.

  Score: 1

7. Command and Control

- **Doctrine.** Battle command doctrine is well-developed—it emphasizes flexibility of command and is based on decentralized command structure with decisionmaking authority delegated to subordinate commanders and staffs. Staffs from the tactical to the strategic levels conduct operations planning based on the command estimate process.

  Score: 3

- **Technology.** Commanders and staffs have relatively modern C4I assets to control the battle (gather information, receive guidance, process and disseminate operational plans).

  Score: 2

- **OR/training.** Country 1's command and control system has been successfully tested in multinational and joint operations, but concerns have been voiced over its ability to withstand exposure to high-intensity conflict.

  Score: 2

8. Communications

- **Doctrine.** Orders formats are based on U.S. doctrine. While most personnel are not fluent in English, Country 1's doctrine calls for the extensive use of well-trained liaisons to interface with U.S. forces.

  Score: 2

- **Technology.** Military architecture for C4ISR includes reconnaissance satellites and battlespace command and control systems for ground forces (the joint aspect is not as well developed). Communications systems are digital and compatible with U.S. systems. While Country 1 has secure communications capabilities, it has only a limited ability for secure communications with U.S. networks.

  Score: 2

- **OR/training.** Country 1 employs a substantial number of liaisons, who are fluent in "U.S. military English" and knowledgeable on U.S. doctrine. Its communication system is fully functional for battlefield operations—reliable and quickly deployable. However, modifications are required for full interoperability with the United States.

  Score: 2

9. Intelligence

- **Doctrine.** Plan to use full range of intelligence capabilities in both joint and combined operations. Well-defined procedures for collection, analysis, and dissemination (e.g., IPB).

  Score: 3

- **Technology.** Broad and sophisticated array of intelligence collection and analysis (HUMNIT, SIGINT, ELINT). Country 1's intelligence sharing network is compatible with the United States; it relies on platforms similar to ASAS Warlord workstations.

  Score: 3

- **OR/training.** Based on recent exercise analysis, Country 1's armed forces are capable of conducting the full range of intelligence operations at the combined arms level. Joint intelligence sharing has only recently been established as a priority, and is therefore lacking.

  Score: 2

Sample MCAT Scorecard Analysis

Figures C.3 and C.4 present the final results of an MCAT analysis. Figure C.3 demonstrates the data as submitted onto the worksheet. From this view, specific weaknesses in CAAs or CMs for a single country become obvious. Note that Country 2 has a critical weakness in deployability compatibility. Country 1, while strong in doctrine, requires some long-term fixes in the operational readiness and training realms.

Manipulation of the MCAT flags creates the MFC scorecard demonstrated in Figure C.4. By listing the countries hierarchically in terms of compatibility, this scorecard presents the data in a form conducive to cross-country comparisons both in overall compatibility and across the compatibility measures of doctrine, technology, and operational readiness and training. Note that the "Compatibility Score" column on the far right of the scorecard shows the average score, while the remaining columns allow for comparison in individual CAAs.

RANDMR1291-C.3

☐ Color Blank Cells

3 High Color Value

0 Low Color Value

COUNTRY	CM	Flag.	BATTLEFIELD FUNCTIONS: Man-euver	BATTLEFIELD FUNCTIONS: Fire Support	BATTLEFIELD FUNCTIONS: Air Defense	BATTLEFIELD FUNCTIONS: Mob/Surv	CSS: Logis-tics	CSS: Deploy-ability	C4I: C2	C4I: Commo	C4I: Intel	Compat-ibility Score Column
	CAAs		1	1	1	1	1	1	1	1	1	Wt or Min → Wt
Country 1	Doctrine	1	3	3	3	2	2	2	3	2	3	2.6
Country 1	Technology	1	2	2	2	2	1	2	2	2	3	2.0
Country 1	OR/TRNG	1	2	2	2	2	1	1	2	2	2	1.8
Country 1	**Overall**	1	2.3	2.3	2.3	2.0	1.3	1.7	2.3	2.0	2.7	2.1
Country 2	Doctrine	1	2	2	1	1	2	1	2	2	1	1.6
Country 2	Technology	1	2	1	1	1	1	0	2	1	1	1.1
Country 2	OR/TRNG	1	2	1	1	1	1	0	2	1	1	1.1
Country 2	**Overall**	1	2.0	1.3	1.0	1.0	1.3	0.3	2.0	1.3	1.0	1.3
Country 3	Doctrine	1	1	1	1	1	2	2	1	2	1	1.3
Country 3	Technology	1	2	2	1	2	1	1	1	2	2	1.6
Country 3	OR/TRNG	1	1	1	1	1	1	1	1	1	1	1.0
Country 3	**Overall**	1	1.3	1.3	1.0	1.3	1.3	1.3	1.0	1.7	1.3	1.3

Figure C.3—Completed MFC Scorecard

RANDMR1291-C.4

☐ Color Blank Cells

3 High Color Value

0 Low Color Value

			BATTLEFIELD FUNCTIONS				CSS		C4I			
		CAAs	Man-euver	Fire Support	Air Defense	Mob/ Surv	Logis-tics	Deploy-ability	C2	Commo	Intel	Wt or Min → Wt
			1	1	1	1	1	1	1	1	1	
COUNTRY	CM	Flag.										Compat-ibility Score Column
Country 1	**Overall**	3	2.3	2.3	2.3	2.0	1.3	1.7	2.3	2.0	2.7	2.1
Country 3	**Overall**	3	1.3	1.3	1.0	1.3	1.3	1.3	1.0	1.7	1.3	1.3
Country 2	**Overall**	3	2.0	1.3	1.0	1.0	1.3	0.3	2.0	1.3	1.0	1.3
Country 1	Doctrine	2	3	3	3	2	2	2	3	2	3	2.6
Country 2	Doctrine	2	2	2	1	1	2	1	2	2	1	1.6
Country 3	Doctrine	2	1	1	1	1	2	2	1	2	1	1.3
Country 1	Technology	1	2	2	2	2	1	2	2	2	3	2.0
Country 3	Technology	1	2	2	1	2	1	1	1	2	2	1.6
Country 2	Technology	1	2	1	1	1	1	0	2	1	1	1.1
Country 1	OR/TRNG	0	2	2	2	2	1	1	2	2	2	1.8
Country 2	OR/TRNG	0	2	1	1	1	1	0	2	1	1	1.1
Country 3	OR/TRNG	0	1	1	1	1	1	1	1	1	1	1.0

Figure C.4—MFC Scorecard Demonstrating Ranking by CMs and Overall Compatibility